The Physics of CT Dosimetry

CTDI and Beyond

Series in Medical Physics and Biomedical Engineering

Series Editors
John G. Webster, Russell Ritenour, Slavik Tabakov, and Kwan-Hoong Ng

Recent books in the series:

Proton Therapy Physics, Second Edition
Harald Paganetti (Ed)

Mixed and Augmented Reality in Medicine
Terry M. Peters, Cristian A. Linte, Ziv Yaniv, Jacqueline Williams (Eds)

Graphics Processing Unit-Based High Performance Computing in Radiation Therapy
Xun Jia, Steve B. Jiang (Eds)

Clinical Radiotherapy Physics with MATLAB: A Problem-Solving Approach
Pavel Dvorak

Advanced and Emerging Technologies in Radiation Oncology Physics
Siyong Kim, John W. Wong (Eds)

Advances in Particle Therapy: A Multidisciplinary Approach
Manjit Dosanjh, Jacques Bernier (Eds)

Radiotherapy and Clinical Radiobiology of Head and Neck Cancer
Loredana G. Marcu, Iuliana Toma-Dasu, Alexandru Dasu, Claes Mercke

Problems and Solutions in Medical Physics: Diagnostic Imaging Physics
Kwan Hoong Ng, Jeannie Hsiu Ding Wong, Geoffrey D. Clarke (Eds)

A Guide to Outcome Modelling in Radiotherapy and Oncology: Listening to the Data
Issam El Naqa (Ed)

For more information about this series, please visit: https://www.crcpress.com/Series-in-Medical-Physics-and-Biomedical-Engineering/book-series/CHMEPHBIOENG

The Physics of CT Dosimetry
CTDI and Beyond

Robert L. Dixon
Wake Forest University School of Medicine

CRC Press
Taylor & Francis Group
Boca Raton London New York

CRC Press is an imprint of the
Taylor & Francis Group, an **informa** business

CRC Press
Taylor & Francis Group
6000 Broken Sound Parkway NW, Suite 300
Boca Raton, FL 33487-2742

First issued in paperback 2021

© 2019 by Taylor & Francis Group, LLC
CRC Press is an imprint of Taylor & Francis Group, an Informa business

No claim to original U.S. Government works

ISBN 13: 978-0-367-78004-3 (pbk)
ISBN 13: 978-0-367-07759-4 (hbk)

Library of Congress Cataloging-in-Publication Data

Names: Dixon, Robert L. (Robert Leland), 1940- author.
Title: The physics of CT dosimetry : CTDI and beyond / Robert L. Dixon.
Other titles: Series in medical physics and biomedical engineering.
Description: Boca Raton, FL : CRC Press, Taylor & Francis Group, [2019] |
Series: Series in medical physics and biomedical engineering | Includes
bibliographical references and index.
Identifiers: LCCN 2018057980| ISBN 9780367077594 (hbk ; alk. paper) | ISBN
0367077590 (hbk ; alk. paper) | ISBN 9780429023330 (eBook) | ISBN
0429023332 (eBook)
Subjects: LCSH: Tomography. | Radiation dosimetry.
Classification: LCC RC78.7.T6 D59 2019 | DDC 616.07/57--dc23
LC record available at https://lccn.loc.gov/2018057980

Visit the Taylor & Francis Web site at
http://www.taylorandfrancis.com

and the CRC Press Web site at
http://www.crcpress.com

Contents

Preface

U NLIKE MOST BOOKS OR textbooks in which CT dosimetry is presented in a "cook-book" format as formulae to be "plugged," in this book derivations and physical rigor are employed throughout in order to impart an in-depth knowledge of the subject to the reader. The results are also accompanied by clear physical descriptions and many illustrative figures for the less mathematically inclined reader, with the more difficult mathematical developments relegated to Appendices. An effort has also been made to keep each chapter self-contained in order to avoid too much page flipping.

Rigorous phantom dose equations have been derived which also illustrate the significant limitations and common misconceptions concerning the CTDI-paradigm. A chapter is devoted to automatic tube current modulation (TCM) and another to stationary patient-support techniques such as perfusion studies using multiple rotations at a fixed z-location and another to wide cone beam techniques in which the desired anatomy can be imaged in a single axial rotation without table motion. Also, analytic equations are derived, based on a simple scatter kernel of Monte Carlo parentage, which strip away the integral facade of the CTDI-paradigm and provide the reader a better physical understanding of CT dosimetry, including the close relationship between stationary and moving table dosimetry. Many dose measurement shortcuts and options are also provided such as use of a small ion chamber and CTDI-aperture. The shortcomings of the scanner-reported $CTDI_{vol}$ are made clear; instructing the reader in detecting and correcting (or ignoring) the resulting anomalous values of $CTDI_{vol}$ and SSDE, with a look to the future in providing more accurate tools for CT dosimetry.

There are also many surprises concerning the subject awaiting the reader. So strap-in and enjoy the flight.

Acknowledgments

DEDICATED TO MY WIFE **Marsha,** who encouraged me in this endeavor as well as during my career.

Also, thanks to:

Professor John Boone of the University of California-Davis who was my principal co-author on many *Medical Physics* papers on CT dosimetry.

And thanks to my other co-authors in that series of papers:

Mike Munley, Robert Kraft, Ersin Bayram, and Adam Ballard – faculty and students of Wake Forest University.

And I am grateful to:

Dr. Shinichiro Mori of The National Institute of Radiological Sciences, Chiba, Japan

for providing to me much of the experimental data used to confirm the theory presented in this book.

And finally, thanks to:

Lt. Bakhit "Quasar" Kourman, Jr. for his invaluable assistance in formatting the book chapters.

Author

ROBERT L. DIXON, PH.D., FACR, FAAPM, is *Professor Emeritus* in the Dept. of Radiology, Wake Forest University School of Medicine and holds a *Ph.D. in Nuclear Physics*.

He has taught physics and medical physics for decades and is a recipient of the Radiology Department's Teaching Excellence Award. He is also a past president of the American Association of Physicists in Medicine (AAPM) as well as past Chairman of the AAPM Science Council and CT Committee; and is a past Vice President of RSNA. He has published many papers on CT dosimetry in the journal *Medical Physics* and is a five-time winner of the SEAAPM Best Publication Award.

Dixon is a member of the IEC MT30 CT committee; he is a designated US CT Expert (ANSI); and has been an invited keynote speaker at multiple international conferences.

He has also been a member of the Governing Board and the Executive Committee of the American Institute of Physics (AIP), and has received the Distinguished Service Award and the Lifetime Achievement Award of the American Board of Radiology (ABR).

He was also an airshow performer (1991–2006) flying former military aircraft including an ex-Soviet Air Force jet.

About the Series

THE *SERIES IN MEDICAL Physics and Biomedical Engineering* describes the applications of physical sciences, engineering, and mathematics in medicine and clinical research. The series seeks (but is not restricted to) publications in the following topics:

- Artificial organs
- Assistive technology
- Bioinformatics
- Bioinstrumentation
- Biomaterials
- Biomechanics
- Biomedical engineering
- Clinical engineering
- Imaging
- Implants
- Medical computing and mathematics
- Medical/surgical devices
- Patient monitoring
- Physiological measurement
- Prosthetics
- Radiation protection, health physics, and dosimetry
- Regulatory issues
- Rehabilitation engineering
- Sports medicine
- Systems physiology

- Telemedicine

- Tissue engineering

- Treatment

THE INTERNATIONAL ORGANIZATION FOR MEDICAL PHYSICS

The International Organization for Medical Physics (IOMP) represents over 18,000 medical physicists worldwide and has a membership of 80 national and 6 regional organizations, together with a number of corporate members. Individual medical physicists of all national member organizations are also automatically members.

The mission of IOMP is to advance medical physics practice worldwide by disseminating scientific and technical information, fostering the educational and professional development of medical physics and promoting the highest quality medical physics services for patients.

A World Congress on Medical Physics and Biomedical Engineering is held every three years in cooperation with International Federation for Medical and Biological Engineering (IFMBE) and International Union for Physics and Engineering Sciences in Medicine (IUPESM). A regionally based international conference, the International Congress of Medical Physics (ICMP) is held between world congresses. IOMP also sponsors international conferences, workshops and courses.

The IOMP has several programmes to assist medical physicists in developing countries. The joint IOMP Library Programme supports 75 active libraries in 43 developing countries, and the Used Equipment Programme coordinates equipment donations. The Travel Assistance Programme provides a limited number of grants to enable physicists to attend the world congresses.

IOMP co-sponsors the *Journal of Applied Clinical Medical Physics*. The IOMP publishes, an electronic bulletin, *Medical Physics World*, twice a year. The IOMP also publishes *e-Zine*, an electronic newsletter about six times a year. The IOMP has an agreement with Taylor & Francis for the publication of the *Medical Physics and Biomedical Engineering* series of textbooks. IOMP members receive a discount.

The IOMP collaborates with international organizations, such as the World Health Organization (WHO), the International Atomic Energy Agency (IAEA) and other international professional bodies such as the International Radiation Protection Association (IRPA) and the International Commission on Radiological Protection (ICRP), to promote the development of medical physics and the safe use of radiation and medical devices.

Guidance on education, training, and professional development of medical physicists is issued by the IOMP, which is collaborating with other professional organizations in development of a professional certification system for medical physicists that can be implemented on a global basis.

The IOMP website (www.iomp.org) contains information on all the activities of the IOMP; policy statements 1 and 2 and the "IOMP: Review and Way Forward" outline all the activities of the IOMP and plans for the future.

CHAPTER 1

Introduction and History

1.1 INTRODUCTION

In most books or textbooks, CT dosimetry is presented in "cookbook" form, namely a 100 mm pencil chamber measurement in a phantom and a formula for $CTDI_{100}$ to be "plugged" (including that of its offspring $CTDI_w$ and $CTDI_{vol}$), without providing a derivation of the formula or discussion of its many limitations. In this book, derivations and physical rigor are employed throughout, making these limitations *and their required corrections* readily apparent to the reader, and made plausible by accompanying the results with clear physical descriptions for the less mathematically inclined reader. An effort has been made to keep each chapter self-contained to avoid too much "page flipping."

1.2 A HISTORICAL VIEW OF CT DOSIMETRY

The following historical vignette lends some perspective on the development of CT dosimetry. This chapter will also serve as an introduction to this book and may not be strictly chronological (and some "literary license" has been employed).

These early workers could not have imagined the explosive growth in CT methodology over the ensuing decades.

1.2.1 The Early Universe

The early measurement of CT dose and mapping of the dose distribution was primarily done using thermoluminescent dosimetry (TLD) which was tedious and had relatively low spatial resolution. In the early days of CT when scan times were slow and x-ray tube heat capacities were low, obtaining the dose (or dose distribution) resulting from multiple axial slices was difficult. Ed McCollough and Tom Payne (beginning in 1976) did some early work using TLD.

In 1977, the pencil chamber method was introduced by Jucius and Kambic – the same year the Apple II computer was released, and people were playing the Atari game, PONG.

Bob Jucius and George Kambic of Ohio Nuclear, Inc. (a US CT manufacturer) provided the first comprehensive look at CT dosimetry, presenting various options including TLD *as well as the introduction of the long pencil ion chamber* which they commissioned Capintec, Inc. to manufacture for them (Jucius et al. 1977). They derived an equation which showed that the integral of a single-slice dose profile could be used to *predict* the average dose about the central scan location ($z = 0$) for multiple slices. This is far from obvious, and their insight was quite impressive. Their derivation involved a (relatively opaque) summation of integrals. They also mapped dose distributions using TLD and surface dose using Kodak RP/M (mammography) film, but concluded that "at this time, TLD is the technique of choice."

Dixon and Ekstrand (1978) independently introduced surface dose mapping using a slower radiation therapy verification film (Kodak Xomat/V), digitized using a scanning densitometer for various scanners of the day (resulting in some unexpected dose spikes).

1.2.2 The Birth of CTDI – 1981

Perhaps the best-known paper was that of a US FDA group – Shope, Gagne, and Johnson (Shope et al. 1981) – who refined the integral concept of Jucius and Kambic described in the previous section. To avoid confusion, we will henceforth adopt the notation used throughout this book. They defined the "multiple slice average dose" (MSAD) resulting from a series of N identical axial dose profiles $f(z)$ spaced at equal intervals of $b = \Delta d$ along z as,

$$MSAD = D_L(0) = \frac{1}{b} \int_{-L/2}^{L/2} f(z')dz' \tag{1.1}$$

where the MSAD is the average dose over $\pm b/2$ about $z = 0$ (at the center of the scan length L) and where $L = Nb$. For axial scans the dose distribution over the scan length is quasi-periodic of period b, hence the average is over one period ($\pm b/2$) about $z = 0$. Note that their nomenclature "multiple scan average dose" (MSAD) is rather misleading, since it is not the average dose over the total scan length, but rather only about the *center of the scan length* $z = 0$. They also stated that L in the above MSAD equation was intended to be long enough for the dose at the center of the scan length to reach its limiting, *equilibrium value*. From this they defined a "dose index" CTDI as,

$$CTDI_\infty = \frac{1}{T} \int_{-\infty}^{\infty} f(z')dz' \tag{1.2}$$

where
$\quad T \quad$ is "the slice thickness as stated by the manufacturer" and
$\quad f(z) \quad$ is the *dose profile* generated by a single axial scan centered at $z = 0$.

$CTDI_\infty$ is the value of MSAD when L is large enough such that MSAD approaches its limiting (equilibrium) value (which we denote by D_{eq}) – such that profiles beyond $z = \pm L/2$ contribute negligible scatter back to $z = 0$; $z = 0$ being the relevant location for MSAD or CTDI. Note also that $CTDI_\infty$ represents the dose that accrues at the center of the scan length for a table increment $b = T$, which represented "contiguous axial scans." With the advent of multi-detector CT (MDCT), T is replaced by "N×T" (nT in our more concise notation used herein). A common misconception is that T or nT represent a beam width, but physically (in any dose formula) they represent a table increment, as will become clear from our derivations in Chapter 2.

The derivation of the MSAD equation by Shope and Gagne (Shope et al. 1981) involved a tedious summation of integrals (following Jucius and Kambic). The derivation for axial scans has been simplified to a few steps (Dixon 2003) using convolution mathematics; this derivation produces the "running mean" dose $D_L(z)$ as an average over $z \pm b/2$ at all values of z (and not just $z = 0$ as for the MSAD of Shope et al.). This derivation is shown in Chapter 2.

1.2.3 Enter the Regulators – 1989

Codification of physical law rarely turns out well, and once the law has been laid down it is devilishly hard to change (or "too many cooks spoil the broth").

The original definition of CTDI put forth by Shope et al. (1981), as well as the original US FDA regulatory proposal (FDA 1984), used the *infinite* line integral of the single-slice, axial dose profile $f(z)$, viz. $L \to \infty$ with $b = T$. The meaning and intent of "infinity" were clear and unambiguous to the physicists, symbolically indicating that the integration limits $(-L/2, L/2)$ must be at least large enough to encompass the complete width of $f(z)$ including its long scatter tails, such that any further increase in L would provide a negligible additional contribution to the accumulated dose at $z = 0$ for a scan length L. This in turn assured that the CTDI, thus defined, would represent the *maximum limiting value* of the accumulated dose at the center of the scan length resulting from multiple, contiguous ($b = T$) scans, namely, the *equilibrium dose* D_{eq}. Had the FDA retained it as originally proposed, it would have been self-correcting and "bulletproof," since many of the ensuing difficulties with CTDI were produced by attempting to define suitable, *finite* integration limits.

But alas, "infinity" did not survive the transformation to the "final FDA rule" (due to public comment; and perhaps because the concept of "infinity" is not in the legal lexicon); and thus the $\pm 7T$ integration limits were adopted – the length of which the FDA stated (1984) "would produce little difference from the originally proposed infinite integral for the largest slices then available" ($T = 10$ mm), and "would be representative of typical clinical scan lengths of $10-15\ T$" (100–150 mm). In hindsight, both conclusions were flawed, and rapid technological advances led to typical body scan lengths of 250 mm or greater. The FDA did, however, retain the required coupling between the integration limits and the divisor T.

1.2.4 The Standard Dosimetry Phantoms

The FDA (1984) defined "standard dosimetry phantom" as a right circular cylinder of polymethyl-methacrylate (PMMA) of diameters of 32 cm (body) and 16 cm body (head) which can accommodate a dosimeter both along its axis of rotation and along a line parallel to the axis of rotation 1.0 cm from its surface. An example of a 32 cm diameter "body phantom" is pictured in Figure 1.1 (albeit longer than the usual 15 cm long "plastic disk").

FIGURE 1.1   A 32 cm diameter CT "body phantom."

Nevertheless, a long period of quiet acceptance prevailed, during which time the mathematical theory behind the pencil chamber and subscripted CTDI methodology was forgotten (many likely had not even seen the derivation) – and some began to believe that they were making an actual "dose" measurement with the pencil chamber. One does not, and cannot, directly measure a dose with a pencil chamber. Not even in air. Among other things, a pencil chamber reading defies the inverse square law $(1/r^2)$. Its reading varies as $1/r$. Many "unwary" diagnostic physicists have fallen into the trap of using the pencil chamber outside of its limited, approved use; supporting the old adage "if the only tool you have is a hammer, you tend to treat everything as if it were a nail." The pencil chamber measures *a dose-integral* in units of *mGy.cm*; so even though your electrometer may read *mGy* (or mR), it is likely not programmed for a pencil chamber (and is actually only measuring the charge collected in Coulombs). See Chapter 3 for pencil chamber calibration methods and units.

1.2.5 Enter $CTDI_{100}$ – 1995

$CTDI_{100}$ (based on a 100 mm long pencil chamber measurement) was introduced (Leitz et al. 1995) around 1995 as a *more practical* indicator of patient dose, and then widely adopted (based on European Commission Study Group 1998). The widespread use of the 100 mm chamber seems to have been an *ad hoc* decision, and not supported by the physics.

The FDA kept the required coupling between the integral divisor and the integration limits; but variable integration limits were not practical for the pencil chamber methodology. However, a fixed integration length can (and does) lead to anomalies.

Since $CTDI_{100}$ has a different value for the central and peripheral phantom axes, a desire to have a single CTDI number (dose index) to represent "dose" for a national survey in Sweden (Leitz et al. 1995) led to an approximate "weighted average" dose across the central scan plane at $z=0$ assuming an *ad hoc* linear variation of $CTDI_{100}$ from the central phantom axis to the peripheral axis namely,

$$CTDI_w = (2/3)CTDI_{100}\,(periphery)+(1/3)CTDI_{100}\,(center) \qquad (1.3)$$

The (1/3, 2/3) weighting proves adequate for $CTDI_{vol}$ (based on $CTDI_{100}$); however, the central axis to peripheral axis dose ratio increases as scan length increases beyond 100 mm due to increased scatter thereon. We also note that the actual dose curve $D(r)$ is not linear, but is sigmoidal, with *zero slope* on the central axis ($r=0$) and again near the phantom surface.

1.2.6 The Advent of Multi-Detector CT (MDCT) – 1998

The divisor of the CTDI integral now becomes nT (or "N × T") which is the *active detector length* as projected back to scanner isocenter and represents the total available scan width for reconstruction. The actual primary beam width (*fwhm*) $a > nT$ is required to keep the penumbra beyond the active detectors, called "over-beaming." MDCT allowed reconstruction of smaller slices than nT but with a concomitant increase in noise, e.g., an acquisition using $nT = 20$ mm, can be reconstructed as four 5 mm slices.

1.2.7 Enter $CTDI_{vol}$ (A Misnomer) but an Improvement since It Eliminates nT (N × T)

$CTDI_w$ was later modified by the IEC (2001) to include the effect of "pitch" (table increment b) on dose as,

$$CTDI_{vol} = p^{-1}\,CTDI_w \qquad (1.4)$$

where $p = b/nT = \Delta d/nT$ applies to both helical and axial scans. The nomenclature $CTDI_{vol}$ is again a misnomer since it does not represent a volume average as its subscript might imply – no average having been taken over the 100 mm scan length; rather it still represents the planar average dose over the central scan plane (at $z=0$) for a 100 mm scan length. Its basis is still $CTDI_{100}$ which is hidden. We also note that nT cancels out in $CTDI_{vol}$ such that only the inverse of the table increment per rotation b^{-1} matters – the divisor nT in $CTDI_{100}$ serves only as a place-keeper.

As the table increment $b \rightarrow 0$, then $CTDI_{vol} \rightarrow \infty$; however, this is nonsensical since the actual dose remains finite. The oft-forgotten required coupling of scan length $L = Nb$ and table increment b in Eq. (1.1) also requires the integration limits to approach zero, resulting in the dose approaching the eminently plausible value $Nf(0)$ where $N =$ number of rotations;

i.e., the N dose profiles $f(z)$ simply pile up on top of each other at $z = 0$, and $CTDI_{vol}$ (calculated from $CTDI_{100}$) no longer has any relevance. This will be shown mathematically in Chapter 5 for stationary table CT, although it is fairly obvious.

1.2.8 Dose Length Product

$DLP = L \times CTDI_{vol}$ is a measure of the total energy deposited in the phantom. Note that DLP does not depend on the scan length L *per se* since $L = Nb$ and $CTDI_{vol}$ is proportional to b^{-1}; thus, b cancels in the product, and DLP really depends only on the number of rotations N or total mAs. Increasing scan length L by increasing pitch alone does not change DLP. Even if the table translation is slowed to a stop ($L \to 0$), DLP remains the same. DLP is by no means *equal* to the total energy deposited since $CTDI_{vol}$ is based on $CTDI_{100}$ – the total energy deposited will be calculated in Chapter 2.

1.2.9 Helical Scanning – Scanning with Continuous Table Motion – 1990

Willi Kalender (Kalender et al. 1989) introduced helical scanning ("spiral CT").

Dixon (2003) derived the equations for helical scanning for the *dose $D_L(z)$* over the entire scan length L, for both the central phantom axis and likewise for the peripheral axis where an angular average over 2π at a fixed value of z is used. This was then shown to reduce to the CTDI-paradigm by setting $z = 0$. This derivation treats the dose rate profile as a traveling wave in the phantom (and is accomplished in a few steps for the central axis on which the dose rate is constant) as shown in Chapter 2.

The same equation for $D_L(z)$ was shown (Dixon 2003) to also apply to axial scanning when a longitudinal "running mean" (average over $z \pm b/2$) is used, which also reduces to the CTDI-paradigm at $z = 0$ as previously discussed. This derivation is likewise shown in Chapter 2 and is easily accomplished using convolution mathematics (as opposed to the tedious summation of integrals previously used by Shope et al. to calculate MSAD and CTDI).

1.3 SLIPPING THE SURLY BONDS OF CTDI

The CTDI-paradigm has many limitations which are not widely appreciated as described in this section. The CTDI-paradigm requires shift-invariance for which no scan (or phantom) parameters can vary with z, therefore it cannot apply to many modern shift-variant CT techniques such as tube current modulation (TCM). It also only applies to phantom-in-motion techniques, and not to stationary patient-support protocols.

1.3.1 An Alternative to the Pencil Chamber – 2003

Dixon in his 2003 paper also described an alternative measurement method to that of the pencil chamber of fixed length which is much more versatile. Unlike early CT scanners, modern CT scanners can scan over any desired length of phantom in a few seconds, therefore integrating the dose from a small ion chamber fixed in the moving phantom can give the accumulated dose for any scan length or clinical protocol, and thus can emulate a pencil chamber of any arbitrary length (and can even be used to measure $CTDI_{100}$). That is, the small ion chamber can be used in this way to create a "virtual pencil chamber" of any desired length. This method has been validated experimentally in detail in Dixon and Ballard (2007) and is described in Chapter 3 where a 0.6cc Farmer ion chamber is shown to give the same result as a 100 mm and 150 mm pencil chamber – and for any other scan length L as well. It is also immune to the shift-variant problems discussed below.

1.3.2 AAPM TG-111 – 2010

A task group of The American Association of Physicists in Medicine published AAPM Report 111 (AAPM 2010) entitled "Comprehensive Methodology for the Evaluation of Radiation Dose in X-ray Computed Tomography" in which the small ion chamber is utilized for measurements rather than the pencil chamber, and which recommends a return to the equilibrium dose D_{eq} as the measurement goal (as originally recommended by Shope et al. 1981 and the FDA). There is no mention in this report of CTDI nor the pencil chamber.

1.3.3 Limitations of the CTDI-Paradigm and the Pencil Chamber Acquisition

The CTDI-paradigm has significant limitations. It only applies to moving patient-support techniques, such as helical scanning, as discussed previously. Every dose profile $f(z)$ in such a scan series must be identical to that integrated by the pencil chamber in order for *the predictive method* of CTDI to be valid; in other words, it requires *shift-invariance* for which no scan parameters can vary with z. That is, it requires constant tube current (mA), constant pitch (or table increment b), and a constant phantom cross-section along z. Therefore, it cannot apply to tube current modulation (TCM) which is commonly utilized today. Dixon and Boone (2014) derive the proper dose equations for such *shift-variant* techniques (TCM and pitch modulation) shown later in Chapters 7 and 8.

The small ion-chamber method has no such restrictions. It can even be deployed in an anthropomorphic phantom. It is measuring an actual dose, and not relying on the predictive methodology of CTDI, which uses the integral of a single scan to *foretell* the dose at the center of the scan length which would accrue if *identical* scans were laid down at equal intervals over a 100 mm scan length as for $CTDI_{100}$ and thence for $CTDI_{vol}$.

1.4 THE IEC ATTEMPTS TO CIRCUMVENT THE LIMITATIONS OF CTDI

If the only tool you have is a hammer, you tend to treat everything as if it were a nail.

$CTDI_{100}$ (thence $CTDI_{vol}$) does in fact have a precise physical meaning: it is equal to the actual accumulated dose in-phantom at the center of a series of contiguous scans ($b = nT$) *covering one specific scan length*, $L = 100$ mm; but it underestimates the limiting equilibrium dose D_{eq} (as well as the accumulated dose for any scan length above 100 mm) – particularly for typical clinical body scan lengths of 250–500 mm which approach the equilibrium dose. It also *overestimates* the dose for $L < 100$ mm.

The IEC (IEC 2016) has attempted to "prop-up" $CTDI_{vol}$ and its "hand maiden," the 100 mm long pencil chamber, in a series of patches. These patches govern the scanner-reported $CTDI_{vol}$, as discussed in the following section.

1.4.1 For Shift-Variant Techniques

For shift-variant techniques such as TCM, the IEC version uses the average of $mA(z)$ *over the entire scan length* as if it were a constant mA in the CTDI-paradigm; whereas $CTDI_{vol}$ *applies only to a 100 mm scan length* – a clear disconnect. This creates a "$CTDI_{vol}$ of the second kind" and the disconnect negates a possible physical interpretation of "$CTDI_{vol}$ (TCM)," as illustrated in Chapter 7. IEC also introduces the absurdities which are supposed to represent local doses: $CTDI_{vol}(z)$ and $CTDI_{vol}(t)$; but which (apart from having units of dose) are not doses at all, but merely surrogates for $mA(z)$ as shown in Chapter 7. The *local dose* at z does not track $mA(z)$ [or $mA(t)$] since it also consists of scatter from the entire scan length. To paraphrase Charles Dickens, local dose also depends on "mA past and mA yet to come."

1.4.2 For the Stationary Phantom/Table

For the stationary phantom/table to which the CTDI-paradigm does not apply, the IEC solution is $CTDI_{vol} = N \times CTDI_w$ where N is the number of rotations. Its failure (and a cure) is illustrated in detail in Chapter 9.

1.4.3 Wide Beam Widths

Another such IEC patch is a response to a paper by John Boone (2007) "The Trouble with $CTDI_{100}$" which illustrates a significant drop-off in the value of $CTDI_{100}$ as the primary beam width becomes comparable to the pencil chamber length ($nT > 40$ mm). This patch is designed to keep $CTDI_{100}$ at the same fraction of $CTDI_\infty$ as that for narrow beams (this fraction being about 0.6 on the central axis of the body phantom). It does so for "phantom-in-motion" scan protocols, but it fails in the realm of stationary phantom dosimetry for which wide cone beams are more commonly used, and for which we provide the *appropriate correction* as shown in Chapter 9.

There is, inexplicably, no patch which provides a correction of $CTDI_{100}$ (thence $CTDI_{vol}$) for scan length using $CTDI_L = H(L)\ CTDI_{100}$ although a plethora of such robust $H(L)$ data exists as described in Chapter 9 as well as in other chapters. This correction would provide an appropriate (albeit approximate) physical interpretation for CTDI (TCM) as illustrated in Chapter 7. In this book, we supply rigorous methods of correcting $CTDI_{vol}$.

1.4.4 Use of the Scanner-Reported CTDI

Despite these differences, CTDI has been widely interpreted and used as an indicator of clinical patient dose by *regulators* and *medical physicists* alike, in *national dose surveys*, in *imaging literature*, *in the clinic*, etc., and *on the CT monitor* for every patient scan.

1.4.5 Size-Specific Dose Estimates (SSDE)

The basic SSDE dose index concept presented in the Report of AAPM Task Group 204 (AAPM 2012) and as revised in AAPM 2014 is an approach to develop a more reasonable *estimate* of patient dose using the scanner-reported $CTDI_{vol}$ and conversion factors that account for differing patient "sizes." In situations where a fixed tube current is employed, and the patient anatomy and circumference are reasonably homogeneous over an entire CT scan, SSDE provides an improved *estimate* of dose as compared to $CTDI_{vol}$. Thus, a small patient will correctly be attributed a relatively higher radiation dose compared to $CTDI_{vol}$ due to reduced attenuation compared to the CTDI phantom.

REFERENCES

AAPM 2010, Report of AAPM Task Group 111, Comprehensive methodology for the evaluation of radiation dose in x-ray computed tomography, American Association of Physicists in Medicine, College Park, MD, (February 2010), http://www.aapm.org/pubs/reports/RPT_111.pdf.

AAPM 2012, Size-Specific Dose Estimates (SSDE) in Pediatric and adult body CT examinations, Report of AAPM Task Group 204, American Association of Physicists in Medicine, (2012).

AAPM 2014, Use of water equivalent diameter for calculating patient size and Size-Specific Dose Estimates (SSDE) in CT, Report of AAPM Task Group 220, The American Association of Physicists in Medicine, (2014).

Boone J.M., The trouble with $CTDI_{100}$. *Med Phys* 34, 1364, (2007).

Dixon R.L., A new look at CT dose measurement: Beyond CTDI. *Med Phys* 30, 1272–1280, (2003).

Dixon R.L., and Ballard A., Experimental validation of a versatile system of CT dosimetry using a conventional ion chamber: Beyond $CTDI_{100}$. *Med Phys* 34(8), 3399–3413, (2007).

Dixon R.L., and Boone J.M., Dose equations for tube current modulation in CT scanning and the interpretation of the associated $CTDI_{vol}$. *Med Phys* 40, 111920 (14pp), (2013).

Dixon R.L., and Boone J.M., Stationary table CT dosimetry and anomalous scanner-reported values of $CTDI_{vol}$. *Med Phys* 41(1), 011907 (5pp), (2014).

Dixon R.L., and Ekstrand K.E., A film dosimetry system for use in computed tomography. *Radiology* 127(1), 255–258, (1978).

Dixon R.L., Munley M.T., and Bayram E., An improved analytical model for CT dose simulation with a new look at the theory of CT dose. *Med Phys* 32, 3712–3728, (2005).

Dixon R.L., Boone J.M., and Kraft R., Dose equations for shift-variant CT acquisition modes using variable pitch, tube current, and aperture, and the meaning of their associated $CTDI_{vol}$. *Med Phys* 4, 111906, (2014). doi: 10.1118/1.4897246.

IEC 60601-2-44, *Medical Electrical Equipment — Part 2-44: Particular requirements for the basic safety and essential performance of X-ray equipment for computed tomography*. International Electrotechnical Commission, Geneva, Switzerland, 2001.

IEC 60601-2-44, *Medical Electrical Equipment — Part 2-44: Particular requirements for the basic safety and essential performance of X-ray equipment for computed tomography*, 3rd ed. International Electrotechnical Commission, Geneva, Switzerland, 2016.

Jucius R., and Kambic G., Radiation dosimetry in computed tomography (CT). *Proc SPIE* 127, 286–295, (1977).

Kalender W.A., Seissler W., and Vock P., Single breath-hold spiral volumetric CT by continuous patient translation and scanner rotation. *Radiology* 173(P), 414, (1989).

Leitz W., Axelson B., and Szendro G., Computed tomography dose assessment: A practical approach. *Radiat Prot Dosim* 57, 377–380, (1995).

Shope T., Gagne R., and Johnson G., A method for describing the doses delivered by transmission x-ray computed tomography. *Med Phys* 8, 488–495, (1981).

U.S. FDA Code of Federal Regulations, Diagnostic x-ray systems and their major components, 21CFR §1020.33, Govt. Printing Office, (August 1984).

CHAPTER 2

Derivation of Dose Equations for Shift-Invariant Techniques and the Physical Interpretation of the CTDI-Paradigm

2.1 INTRODUCTION

The basic pencil chamber measurement concept was introduced four decades ago by Bob Jucius and George Kambic of Ohio Nuclear/Technicare (Jucius and Kambic 1977), and the CTDI-paradigm was formally developed by a group from the US FDA (Shope et al. 1981).

Unfortunately, the CTDI-paradigm is a complex concept – and not just a formula to be plugged as shown in Eq. (2.1) below,

$$CTDI_{100} = \frac{1}{nT} \int_{-50\,\text{mm}}^{50\,\text{mm}} f(z')dz' \qquad (2.1)$$

where $f(z)$ is the in-phantom dose distribution resulting from a single axial rotation; with the integral typically being directly acquired using a 100 mm long pencil ionization chamber (in which nT is sometimes written as "N×T").

There is nothing straightforward or obvious about it.

In this book, unless otherwise noted, "the phantom" refers to the standard dosimetry body phantom – a 32 cm diameter PMMA (acrylic) cylinder with its "peripheral axes" located 1 cm below its surface. A glossary of parameters has been appended for quick reference.

Its original derivation for axial scanning (Shope et al. 1981) involved a complex and tedious summation of integrals which has only recently been simplified (Dixon 2003) using convolution mathematics as illustrated later in this chapter. This formula has significant limitations with respect to modern CT techniques such as tube current modulation (TCM) and stationary table scans, and these shortcomings will be revealed in this chapter by a rigorous treatment of the physics.

The best (and only) way to understand the CTDI-paradigm and its many limitations is to derive it as illustrated in the next section.

2.2 DERIVATION OF THE DOSE EQUATIONS AND THE CTDI-PARADIGM ON THE PHANTOM CENTRAL AXIS FOR A *SHIFT-INVARIANT* HELICAL TECHNIQUE IN WHICH NO PARAMETERS VARY WITH Z (CONSTANT TUBE CuRRENT, PITCH, APERTURE, ETC.)

It is relatively simple to derive the equations of the CTDI-paradigm (Dixon 2003) for the accumulated dose $D_L(z)$ on *the phantom central axis* for this case. Translation of the table and phantom at velocity υ produces a constant *dose rate* profile on the phantom central axis in the form of a traveling wave $\dot{f}(z-\upsilon t) = \tau^{-1} f(z-\upsilon t)$ as depicted in Figure 2.1, where $f(z)$ is the single-rotation (axial) dose profile acquired with the phantom held stationary, and τ is the gantry rotation period (in seconds).

The profile shown in Figure 2.1 has been generated by a primary beam width of only 26 mm, but its scatter tails extend over more than 300 mm. The dose accumulated at a fixed value of z (depicted in Figure 2.1) as the profile travels by, is given by the time-integral of $\dot{f}(z-\upsilon t)$ over the total "beam-on" time t_0, namely,

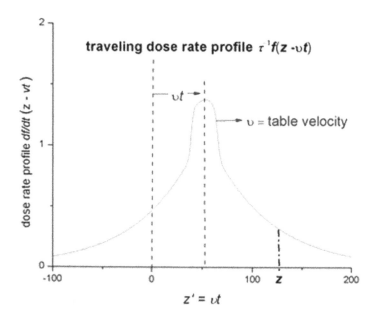

FIGURE 2.1 A traveling *dose rate profile* $\dot{f}(z-\upsilon t) = \tau^{-1} f(z-\upsilon t)$ in the phantom reference frame is created when an axial dose profile $f(z)$ is translated along the phantom central axis z by table translation at velocity υ, where τ is the gantry rotation period (in sec), which has the familiar form of a traveling wave.

(Reprinted from Dixon and Boone, *Medical Physics*, 2013.)

$$D_L(z) = \tau^{-1} \int\limits_{-t_0/2}^{t_0/2} f(z - \upsilon t)\,dt = \frac{1}{b}\int\limits_{-L/2}^{L/2} f(z - z')\,dz' = \frac{1}{b}f(z) \otimes \Pi(z/L) \qquad (2.2)$$

the conversion from the temporal to the spatial domain having been made using $z' = \upsilon t$, scan length $L = \upsilon t_0$, and a table advance per rotation $b = \upsilon\tau$, resulting in the above convolution equation describing the total dose $D_L(z)$ accumulated at any given z-value during the complete scan. $\Pi(z/L)$ represents a rectangular function of unit height and length L. The reader will note the long scatter tails on the dose profile in Figure 2.1 such that the point z will begin accumulating dose long before the primary beam component has arrived and long after it has passed. The convolution process is depicted schematically in Figure 2.2.

The accumulated dose at the center of the scan length $(-L/2, L/2)$ is easily obtained by setting $z = 0$ in Eq. (2.2), namely,

$$D_L(0) = \frac{1}{b}\int\limits_{-L/2}^{L/2} f(z')\,dz' \qquad (2.3)$$

The above equations have likewise been shown to be valid (Dixon 2003) for axial scanning using the "running mean" (an average over $z \pm b/2$), as well as for helical scans on the peripheral axes using an angular average over 2π at a fixed z as rigorously derived later in this chapter.

The resemblance of Eq. (2.3) to the CTDI equation is obvious. By setting the table increment to $b = nT$ (a helical pitch $p = b/nT$ of *unity* or an axial scan interval $b = \Delta d = nT$) and, by arbitrarily truncating the scan length to $L = 100$ mm, Eq. (2.3) reduces to Eq. (2.1)

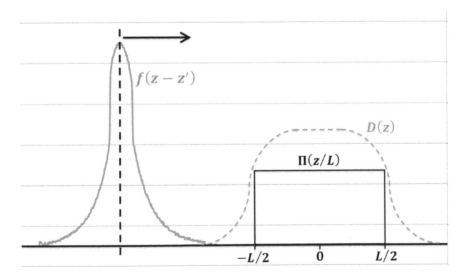

FIGURE 2.2 Schematic graphical depiction of the convolution in Eq. (2.2). The rectangular function can be considered as the normalized tube current (mA) profile.

(From Dixon, 2018.)

for $CTDI_{100}$. Thus, $CTDI_{100}$ is equal to the dose at the center of the scan length $z = 0$ for a 100 mm scan length for a table increment per rotation of $b = \Delta d = nT$.

A separate stand-alone equation for $CTDI_L$ is unnecessary and redundant, since $CTDI_L = pD_L(0)$. Nonetheless, it is included below for future reference.

$$\mathrm{CTDI}_L = \frac{1}{nT} \int\limits_{-L/2}^{L/2} f(z')dz' \tag{2.4}$$

Since $CTDI_{100}$ has a different value for the central and peripheral phantom axes, a desire to have a single CTDI number to represent "dose" led to an approximate "weighted average" dose across the central scan plane at $z = 0$ (Leitz et al. 1995), namely,

$$CTDI_w = (2/3)CTDI_{100}(\text{periphery}) + (1/3)CTDI_{100}(\text{center}) \tag{2.5}$$

This was later modified by the IEC to include the effect of "pitch" (table increment b) on dose as,

$$CTDI_{\mathrm{vol}} = p^{-1}CTDI_w \tag{2.6}$$

where $p = b/nT = \Delta d/nT$ applies to both helical and axial scans. The nomenclature $CTDI_{\mathrm{vol}}$ is a *misnomer* since it does not represent a volume average as its subscript might imply – no average having been taken over the 100 mm scan length; rather it still represents the planar average dose over the central scan plane (at $z = 0$) for a 100 mm scan length (its basis is still $CTDI_{100}$). There is a misconception that $CTDI_{100}$ represents the average dose over 100 mm since it is based on a 100 mm long integral as acquired by a pencil ionization chamber of the same length. This is incorrect as can be seen from the derivation. The CTDI-paradigm is predictive only – predicting the accumulated dose at the center of the scan length which would accrue if a 100 mm long scan series were performed. Using the measurement of the integral of a dose profile over $(-L/2, L/2)$ from a single, axial rotation, it allows one to "fore-tell" the dose $D_L(0)$ which would occur at $z = 0$ for multiple rotations evenly spaced over the same phantom length L, where each rotation delivers the same mAs used for the integral measurement.

It is also important to note that the divisor nT in the CTDI equation physically represents a table increment and *not* a beam width, thus the CTDI-paradigm does not apply to stationary table techniques (Dixon et al. 2014). For multi-detector CT (MDCT), the actual primary beam width a (*fwhm*) at isocenter must be larger than nT where nT is the active detector length as projected back to isocenter ($a > nT$ is called "over-beaming"). This is required to keep the penumbra beyond the active detectors on either end. Since $a > nT$ (by more than a factor of 2 for narrow beam widths), calling nT the "nominal beam width" is both misleading and non-scientific terminology. The primary beam *fwhm* a is equal to the z-collimator aperture projected to isocenter (in the usual case where a is greater than the penumbra width generated by the focal spot).

So how does a detector length nT arise as a parameter in a dose equation? From the rather arbitrary (but logical) definition of "contiguous" axial scans $b = \Delta d = nT$ and a helical pitch of unity $p = b/nT = 1$. That is, nT is the maximum available reconstructed slice width for axial scans.

2.3 LIMITATIONS OF THE CTDI-PARADIGM: THE REQUIREMENT FOR *SHIFT-INVARIANCE*

We also note from the derivation that there are significant restrictions which apply to the CTDI-paradigm – namely, it requires *shift invariance* along z. That is, no scanner (or phantom) parameters can change along z; hence it applies only for constant tube current (mA); constant pitch; and constant z-collimator aperture a; as well as constant phantom dimensions along z (e.g., it could also apply to an elliptical phantom of constant cross-section). It is also important to note that the divisor b and the integration limits $\pm\,L/2$ are *coupled* via $L = Nb$ where $N =$ number of rotations, thus implicit in the $CTDI_{100}$ formula is the often overlooked requirement for $N = 100/nT$ rotations, and the fact that nT physically represents a particular value of table increment per rotation and not a beam width – "nominal" or otherwise. For example, for $nT = 5$ mm, the actual beam width a may be 7–8 mm, and the $CTDI_{100}$ value represents the central ($z = 0$) dose for 20 rotations spaced at 5 mm intervals. If $N < 20$, then the predictive value of $CTDI_{100}$ is lost, since the derivations assume scatter is present for 20 such rotations.

The necessity for *shift-invariance* in CTDI can be also be operationally understood from the nature of a pencil chamber acquisition; namely, *all profiles in the scan series must be identical to the single profile integrated by the pencil chamber* in order for the equations and predictive nature of the CTDI-paradigm to apply.

On the other hand, the direct measurement method using a small, Farmer-type ionization chamber (Dixon 2003; Dixon and Ballard 2007), and as recommended by AAPM Task Group 111 (AAPM 2010) is unaffected by *shift-variance*. This method will be covered in detail in Chapter 3.

That notwithstanding, the IEC (2016) has felt an imperative to report a CTDI value for any *shift-variant* scan technique such as tube current modulation (TCM), and also including stationary phantom techniques. These *ad hoc* definitions and their shortcomings will be described later.

Integral dose (total energy deposited) and *DLP* are typically immune to *shift-variance*.

Figure 2.3 illustrates the use of the convolution in Eq. (2.2) to generate the accumulated dose distribution (Dixon et al. 2014).

We note that $D_L(0) = 5.0$ gives the peak dose for a constant mA (*shift invariant*) dose distribution. As the scan length increases further beyond $L = 276$ mm, the dose approaches a limiting equilibrium value of $D_{eq} = 5.4$ corresponding to infinite integration limits in Eqs (2.2–2.4). The approach to equilibrium is asymptotic and for practical purposes is reached (to within 2%) when $L \geq L_{eq} = 470$ mm.

The limiting equilibrium dose $D_{eq} = 5.4$ is first approached at the center ($z = 0$) for $L \geq 470$ mm and then spreads over a wider range of z as L is further increased (analogous to inflating a balloon against a flat ceiling). The average dose over the scan length $\bar{D}_L$ is

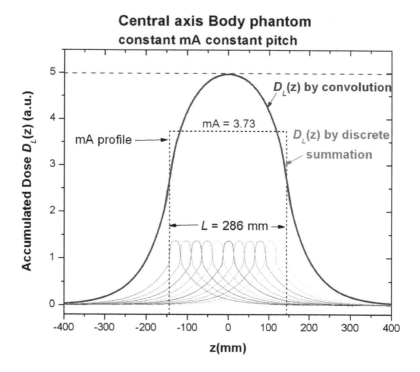

FIGURE 2.3   Accumulated dose at constant mA using the convolution Eq. (2.2) for $D_L(z)$ and also using the discrete superposition (summation) of the 11 dose profiles depicted, each having an aperture of $a = 26$ mm and spaced at like intervals using a table increment $b = a = 26$ mm (no primary beam overlap) – the discrete summation distribution being essentially indistinguishable from the convolution in this case.

(Reprinted from Dixon and Boone, *Medical Physics*, 2013.)

typically about 10% below the peak dose for clinically relevant scan lengths (Dixon and Boone 2013), $\bar{D}_L = 0.88\ D_L(0) = 4.4$ in this case. For an $L = 100$ mm scan length, only 56% of the total energy E deposited is deposited by scatter and primary radiation within the directly irradiated $L = 100$ mm interval $(-L/2, L/2)$ – the remaining 44% is deposited by scatter alone outside the interval $(-L/2, L/2)$.

It is easy to visualize from the convolution picture (Figure 2.2) how the accumulated dose profile builds up as the single-slice profile is slid into the box $(-L/2, L/2)$ and the product integrated. It is clear that the dose in the central region will flatten out and reach an equilibrium value when L is large enough to encompass essentially all of the scatter tails of $f(z)$, i.e., when the trailing scatter tail of $f(z)$ is totally inside the box. This equilibrium dose value is given by,

$$D_{eq} = \frac{1}{b} \int_{-\infty}^{\infty} f(z)\,dz \qquad (2.7)$$

which corresponds to $CTDI_\infty$ for a table increment of $b = nT$.

The more realistic dose profile illustrated in Figure 2.1 corresponds to a primary beam width of only 26 mm, yet the scatter tails extend over more than 400 mm. A convolution box $\Pi(z/L)$ with $L = 470$ mm is required to encompass most of it and thus approach the equilibrium dose on the central axis of the body phantom. As seen in Figure 2.3, for realistic beam profiles $f(z)$, the accumulated dose along the axis $D_L(z)$ is quite non-uniform, varying by a factor of two over the scan length; whereas it is the peak dose which is predicted by the CTDI-paradigm in this case; namely $D_L(0) = p^{-1} CTDI_L$, where p is the helical pitch $p = b/nT$.

Since $b = \upsilon\tau$ is the table advance per rotation, we note that the faster the table is translated, the lower the dose, which is to be expected since the same amount of radiation (energy) is spread over a larger distance.

Whether or not dose equilibrium is reached depends only on the relative width of the dose profile $f(z)$ (including its "scatter tails") and the total scan length L, and does not depend on couch speed υ (or pitch $\upsilon\tau/nT$).

The equations derived above are consistent with previous empirical results for helical scans obtained using TLD measurements (McNitt-Gray and Cagnon 1999; McGee and Humphreys 1994). Note that the proper value by which to divide $\int_{-\infty}^{\infty} f(z)dz$ to obtain an actual accumulated dose is the table advance per rotation $b = \upsilon\tau$ and not the value of nT as with CTDI.

The CTDI can be a confusing concept since it is not an actual dose, but an expectation (Shope et al. 1981; FDA 1984). The actual measured quantity is the dose integral, and the divisor nT has nothing to do with the dose, but merely acts as a place-keeper (and a reminder of the collimator configuration used in the measurement of the dose integral). Also note that nT cancels out in $CTDI_{vol} = (nT/b) CTDI_w$ such that $CTDI_{vol}$ depends only on the inverse of the table increment per rotation b (or inverse pitch).

Failure to recognize the divisor nT in the CTDI equation *as a table increment* can lead to misapplication of CTDI to situations where it has no validity, such as stationary phantom CT (SCBCT) for which $b = 0$. Without table/phantom advance, $b = 0$ and $L = Nb = 0$; thus there are no "contiguous" scans ($b = nT$ has no meaning and pitch $= 0$); therefore nT has no relevance (nor should it even appear) in any valid dose equation; and it is clear that Eq. (2.4) for $CTDI_L$ cannot apply (nor can it be derived) for the stationary phantom.

2.4 EXTENSION OF THE DERIVATIONS TO AXIAL SCANS AND TO HELICAL SCANS ON THE PERIPHERAL AXES

Eqs (2.2–2.4) have likewise been shown (Dixon 2003) to apply to helical scanning on a peripheral axis at a pitch $p = b/nT$ if an *angular average* over 2π *at a fixed z* is used to smooth (average) the peripheral axis dose distribution rather than the longitudinal, running-mean used for axial scans [the dose for helical scans on the phantom central axis is non-oscillatory, requiring no averaging, and is given by Eq. (2.2)]. The longitudinal and angular averages have been shown (Dixon 2003) to converge at values of z where dose equilibrium has been established.

2.4.1 Derivation of the Dose Equations for Axial Scans

It is instructive to derive (Dixon 2003) similar equations for axial scanning using the same formalism, both for completeness and also because the equations describe the mean dose distribution as a continuous function of z; whereas the MSAD and dose index (CTDI) derived by Shope et al. and the FDA give only the mean dose only at the center of the scan length ($z = 0$). In addition, the following derivation using the convolution is much simpler and easier to understand than the tedious sum of integral methods previously utilized (Jucius and Kambic 1977; Shope et al. 1981); and the dose along the peripheral axes in helical CT is a somewhat similar problem.

In the case of axial scans, a single rotation is made at each of a series of equally spaced locations along the z-axis, with no phantom motion during the beam-on time. If we assume that the spacing between scan centers (scan interval) is b and that $N = (2J + 1)$ total scans are utilized with the center scan located at $z = 0$, then the cumulative dose is given by,

$$D(z) = \sum_{n=-J}^{J} f(z - nb) = f(z) \otimes \sum_{-J}^{J} \delta(z - nb) \tag{2.8}$$

where $f(z - nb)$ has been written as the convolution of $f(z)$ with a finite "comb" of δ – functions which serve to *replicate* $f(z)$ at each location $z' = nb$. The cumulative dose $D(z)$ is quasi-periodic of period b. Note that the convolution operation is commutative, associative, and distributive (Bracewell 2000).

The periodicity of $D(z)$ can be averaged over the peaks and valleys by computing the running mean over one period b, which can also be expressed as a convolution,

$$\bar{D}(z) = \frac{1}{b} \int_{z-b/2}^{z+b/2} D(z')dz' = D(z) \otimes \frac{1}{b} \Pi(z/b) \tag{2.9}$$

combining Eqs (2.8 and 2.9), and utilizing the associative property of the convolution, the running mean dose is therefore,

$$\bar{D}(z) = \frac{1}{b} f(z) \otimes \sum_{n=-J}^{J} \delta(z - nb) \otimes \Pi(z/b) = \frac{1}{b} f(z) \otimes \sum_{-J}^{J} \Pi\left(\frac{z - nb}{b}\right) \tag{2.10}$$

The summation in Eq. (2.10) is just a string of rectangular functions of width b laid down edge to edge along the z-axis, and is therefore equivalent to one long rectangular function of length $L = (2J + 1)b = Nb$, centered at $z = 0$, namely $\Pi(z/L)$. Therefore Eq. (2.10) becomes,

$$\bar{D}(z) = D_L(z) = \frac{1}{b} f(z) \otimes \Pi(z/L) = \frac{1}{b} \int_{-L/2}^{L/2} f(z - z')dz'. \tag{2.11}$$

Thus, the equation for the running mean dose in axial scanning can be seen to have the same form as Eq. (2.2) for helical scanning if one replaces $\upsilon\tau$ with b and notes that the "scan length" L is defined as Nb in both cases, where N is the total number of rotations. Thus, we can visualize the running mean dose in axial scanning being created with the same convolution picture and shape as previously depicted in Figure 2.3 for helical scanning. Evaluation of Eq. (2.11) at $z=0$ likewise results in the accumulated dose $D_L(0)$ *at the center of the scan length L* as given previously by Eq. (2.3) and shown again below,

$$D_L(0) = \frac{1}{b} \int_{-L/2}^{L/2} f(z')dz' \qquad (2.12)$$

which (for axial scans) represents an average dose over the small interval $\pm b/2$ about $z=0$, where b is typically small compared to the total scan length $L=Nb$. Note also the implicit dependence of the integration limits $\pm L/2$ and the divisor b, coupled by $L=Nb$. We also note that $CTDI_L = (b/nT)D_L(0)$, so these equations for axial scanning are the same as those previously derived for helical scanning.

The mean equilibrium dose at the center of the scan length is likewise given by,

$$\bar{D}_{eq}(0) = \frac{1}{b} \int_{-\infty}^{\infty} f(z)dz = MSAD_{eq} \qquad (2.13)$$

We again emphasize that the relevant parameter with which to divide $\int_{-\infty}^{\infty} f(z)dz$ to get a real accumulated dose is the scan interval b and not the value of nT as with $CTDI_L$.

2.4.2 Derivation of the Helical Dose Distribution on the Peripheral Axes

For points located off the axis of rotation, such as the peripheral dosimetry axes of the CDRH dosimetry phantom located near the surface, the dose rate is not constant over the rotational period due primarily to varying x-ray attenuation and secondarily to beam divergence (inverse square law), such that the greatest dose is delivered on the entrant side when the x-ray tube (gantry) position is in the same angular quadrant as the dose point (axis). For axial scans, this is also the case, and in both cases the single-slice beam profile $f(z)$ is taken as the average (angular integral) of the dose rate profile $\dot{f}(z,\theta)$ over a single rotation of 360 degrees *with no phantom motion*.

For helical scans, the time variation of dose rate will result in the cumulative dose $D_L(z)$ being quasi-periodic of period $\upsilon\tau$ along a longitudinal, peripheral axis, somewhat like that observed in axial scans, so we must look for some similar averaging method. It is clearly important to be able to relate the measured dose quantity on a peripheral axis $\int_{-\infty}^{\infty} f(z)dz$ (or $CTDI_{100}$) resulting from a single axial rotation to an average helical dose of some kind, as was done for axial scans (using the running mean). In order to visualize the peripheral dose distribution, Xomat/V film was wrapped over the top part of the 32 cm diameter body phantom, covering about 150 degrees of arc on the phantom. Figure 2.4 illustrates the surface dose distribution delivered on the GE Lightspeed, multi-slice scanner for a $nT = 4 \times 3.75$ mm detector configuration with $\upsilon\tau = 22.5$ mm/rotation (HS mode, pitch = 1.5). For the HQ scan mode, 4×5 mm detector configuration, 15 mm/rotation (pitch = 0.75), the surface "gaps" were reduced to about 1 mm.

The dose profile at the phantom entrant surface is narrower than that at isocenter due to beam divergence from the x-ray source. If M = relative magnification factor from the phantom surface to isocenter produced by the divergence of the beam, i.e., $M = S/(S-R)$ where S is the source to isocenter distance and R is the phantom radius; for a pitch equal to $1/M$, the gaps in the surface dose would be minimal, and the peripheral dose relatively smooth. For the GE Lightspeed utilized, $S = 541$ mm, thus $M = 1.4$ for the body phantom which corresponds to a "smoothing" pitch of $1/M = 0.70$. For the head phantom $M = 1.2$, and $1/M = 0.85$.

The *instantaneous dose rate* on a peripheral axis located at angle θ_z is periodic of period τ, and can be represented as a function of x-ray tube (gantry) angle $\theta = \omega t$ ($-\pi \leq \theta \leq \pi$) where $\omega = 2\pi/\tau$, as,

$$\dot{f}(z,t) = \frac{1}{\tau} f(z, \theta - \theta_z) \tag{2.14}$$

the dose rate being maximum when $\theta = \theta_z$.

The axial dose profile $f(z)$ is the integral of Eq. (2.14) over a complete rotation ($\omega\tau = 2\pi$) at a fixed z (no phantom translation), viz.,

$$f(z) = \frac{1}{2\pi} \int_{\theta_0}^{\theta_0 + 2\pi} f(z, \theta - \theta_z) d\theta = \frac{1}{2\pi} \int_{\phi_0}^{\phi_0 + 2\pi} f(z, \theta - \theta_z) d\theta_z \tag{2.15}$$

where θ_0 is any arbitrary angle (as long as the integration range is 2π, the result is the same). Also note that the value of the integral is the same whether the integration is taken over θ or θ_z (whether we choose to rotate the gantry or the phantom – the result is the same).

The traveling dose rate profile with a phantom velocity υ is given by,

$$\dot{f}(z - \upsilon t, t) = \frac{1}{\tau} f(z - \upsilon t, \theta - \theta_z) \tag{2.16}$$

Integrating the dose rate in Eq. (2.16) over the total beam-on time ($-t_0/2, t_0/2$) with $z' = \upsilon t$, $L = \upsilon t_0$, and $\theta = \omega t$, the cumulative dose along the peripheral axis located at angle θ_z is

$$D_L(z, \theta_z) = \frac{1}{\upsilon\tau} \int_{-L/2}^{L/2} f(z - z', \theta - \theta_z) dz' \tag{2.17}$$

Note that $z' = \upsilon t$ and $\theta = \omega t$ are *not* independent variables, and since both the magnitude and shape of $f(z, \theta)$ change with θ, separation of variables is not possible.

Two different methods for averaging this quasi-periodic dose are possible. An "angular average" of the dose over all peripheral axes θ_z from 0 to 2π; or, the "running mean" dose over z as utilized in Section 2.5 for axial scans. The angular average corresponds to an average in the horizontal direction in Figure 2.4 (i.e., rotating the phantom by 360° and smearing out the dose distribution), and the running mean to an average in the vertical z direction in Figure 2.4.

The *angular average* dose is obtained by averaging Eq. (2.17) over all axes θ_z,

$$\bar{D}_{ang}(z) = \frac{1}{\upsilon\tau} \int\limits_{-L/2}^{L/2} \left\{ \frac{1}{2\pi} \int\limits_{0}^{2\pi} f(z-z',\theta-\theta_z)d\theta_z \right\} dz' \qquad (2.18)$$

The inner integral over θ_z in brackets, performed at a fixed value of z' and θ, will be recognized from Eq. (2.16) as the axial dose profile $f(z-z')$, hence the angular average dose is,

$$\bar{D}_{ang}(z) = \frac{1}{\upsilon\tau} \int\limits_{-L/2}^{L/2} f(z-z')dz' = f(z) \otimes \Pi(z/L) \qquad (2.19)$$

which has the same form as the helical dose distribution on the central axis, Eq. (2.2), hence all of the equations previously derived for the helical scan on the central axis [Eqs (2.1–2.4)] apply equally well to the angular average dose on a peripheral axis. It is the angular average of the accumulated dose which is generated by the integral of the axial dose profile $f(z)$ in helical CT. We note that $b = \upsilon\tau$ is the table increment per rotation as before.

2.4.3 Longitudinal Average vs. Angular Average for Helical Scans

Since the running mean (a longitudinal average over $z \pm b/2$) is used in axial CT scanning (e.g., in calculating MSAD about $z = 0$), it behooves us to investigate the relationship of this longitudinal average to the angular average on a peripheral axis. Dixon has shown (2003) that for helical scans on a peripheral axis, the angular and longitudinal (*running mean*) averages are different but converge in regions near $z = 0$ where dose equilibrium has been attained. Dose equilibrium at the center of the scan length on the peripheral axis of the body phantom is achieved for scan lengths $L > 300$ mm. There is a simpler derivation of this convergence at equilibrium than the mathematical proof presented in Dixon (2003) which is given in Chapter 4.

We also remind the reader that the equilibrium dose D_{eq} and $D_L(0) = p^{-1}CTDI_L$ for axial scans, and also for helical scans *on the peripheral axis, refer to average doses* at the center of the scan length, and hence the actual doses in these cases may be quasi-periodic (with peaks and valleys); although the oscillations at equilibrium will be somewhat damped by increased scatter vs. primary. In the case of axial scans, the oscillations are smoothed at a table increment $b = a$ where a is the z-collimator aperture (the primary beam *fwhm*) such that there is no primary beam overlap (or gaps) – the primary beam being responsible for the oscillations. This was, in fact, illustrated in Figure 2.3 which included a superposition of axial scans having $b = a$ which produced a smooth dose distribution essentially indistinguishable from that of the convolution of Eq. (2.2). Figure 2.4 gives a visual depiction of the radiation "stripes" that would be "painted" on the surface of a 32 cm diameter patient at a

FIGURE 2.4   Film image of the dose distribution at the surface of the body phantom for a helical scan of pitch = 1.5 with a 4×3.75 mm acquisition (HS mode) on a GE Lightspeed scanner.

(Reprinted from Dixon, *Medical Physics*, 2003.)

large pitch $p = 1.5$, whereas the CTDI-paradigm and convolution only predict the average over one period. Thus, while a larger pitch reduces the average dose it may mask higher surface dose excursions.

2.5 TOTAL ENERGY E ABSORBED IN THE PHANTOM (AND DLP)

Using the convolution format of $D_L(z)$,

$$D_L(z) = \frac{1}{b} f(z) \otimes \Pi(z/L) = \frac{1}{b} \int_{-L/2}^{L/2} f(z-z')dz' \tag{2.20}$$

It is straightforward to show from the properties of the convolution shown below,

$$\int_{-\infty}^{\infty} \left[f(z) \otimes g(z) \right] dz = \int_{-\infty}^{\infty} f(z)dz \int_{-\infty}^{\infty} g(z)dz \tag{2.21}$$

that the total energy E absorbed in the phantom along (and about) a given z-axis is given by,

$$E = \int_{-\infty}^{\infty} \tilde{D}_L(z)dz = N \int_{-\infty}^{\infty} f(z)dz = LD_{eq} \tag{2.22}$$

where $N =$ total number of rotations and $L = Nb$. The resemblance of E to DLP is obvious except that $\mathrm{DLP} = LCTDI_{vol}$ and thus is based on $CTDI_{100}$, therefore $\mathrm{DLP} < E$ but serves as a surrogate to E.

As will be shown in the chapter on *shift-variant* techniques such as variable mA (TCM) and variable pitch, E and *DLP* are robust (invariant) with respect to *shift-variant* techniques, whereas $CTDI_{vol}$ is not. We also note that neither E nor DLP depend on the scan length *per se* despite its appearance in their respective equations but rather E (and DLP) depend only on the *total mAs* and the z-collimator aperture a (a = primary beam *fwhm*). That is, since $L = Nb$ and both D_{eq} and $CTDI_{vol}$ depend on b^{-1} [Eq. (2.8)]; the table increment b cancels in the product. *Therefore, for a given kV and beam filter, the total energy absorbed (integral dose) E (and its surrogate DLP) depend only on the product of total mAs and z-collimator aperture a.*

So that's it – *E and DLP depend only on* $total\,mAs = \int i(t)dt = \langle i \rangle t_0$, *and collimator aperture a*, and are indifferent as to how the N rotations are spread out along the z-axis (Dixon and Boone 2013); in fact E and DLP remain unchanged even if the table should stop moving ($b = 0$ and $L = Nb = 0$). However, for a given E, *the accumulated dose* $D_L(z)$ *will depend profoundly on how the energy E is spatially distributed along z* (on E per unit length); depending on L and the functional form of $i(z)$ for a TCM protocol.

The collimator aperture a has a more fundamental dosimetric role to play beyond just axial smoothing when the table increment $b = a$, where a is also the primary beam width (*fwhm*) for all values of interest in MDCT.

For MDCT, $a > nT$ to keep the primary beam penumbra beyond the active detector width nT ("over-beaming"). For narrow beam widths, a may exceed nT by more than a factor of two.

The "dose efficiency" is given by (nT/a) as a percentage.

Further, it will be shown in the chapter on analytic equations (Chapter 6) that,

$$D_{eq} = \left(\frac{a}{b}\right) f_p(0)[1+\eta] = \frac{1}{b}\int_{-\infty}^{\infty} f(z)dz \tag{2.23}$$

where $f_p(0)$ is the primary beam intensity on the axis, and η is the scatter-to-primary ratio ($\eta = 13$ on the central axis of the body phantom).

Thus, from $E = LD_{eq}$ and $L = Nb$,

$$E = Naf_p(0)[1+\eta] \tag{2.24}$$

such that the energy deposited E is proportional to the number of rotations N and the aperture a, logical since the aperture a determines the amount of primary beam energy escaping the collimator and the product of N with the mAs per revolution is the total mAs. This also applies to DLP.

2.5.1 CTDI-Aperture

Since $CTDI_L$ will be different for each value of the divisor nT, Dixon et al. (2005) introduced the concept of CTDI-aperture shown in Eq. (2.25), which remains constant for all values of the aperture a and their corresponding values of nT,

$$CTDI_a = \frac{1}{a}\int_{-\infty}^{\infty} f(z)dz \tag{2.25}$$

Although this is a form of $CTDI_\infty$, the value remains remarkably constant even for $CTDI_{100}$ as will be illustrated in Chapter 3. This constancy can be exploited to reduce the measurement burden on the medical physicist.

2.5.2 The Physical Meaning of $CTDI_{\text{free-in-air}}$

This also begs the question, what is the physical meaning of $CTDI_{\text{free-in-air}}$ as measured by the pencil chamber under scatter-free conditions, which also uses the divisor $nT=$"$N \times T$." Is it actually a bona-fide "dose" (air kerma)? The pencil chamber collects the infinite integral of the primary beam profile $f_p(z)$, and since there is no phantom scatter, its peak height $f_p(0)$ (the "dose") does not vary with the primary beam width (aperture) a; however, the infinite integral acquired by the pencil chamber increases linearly with a. Therefore, the infinite integral is the product of $f_p(0)$ and a and thus CTDI-aperture in Eq. (2.25) gives the actual dose $f_p(0)$; however, $CTDI_{\text{free-in-air}}$ results in a value of $(a/nT) f_p(0)$ which is greater than the actual dose since $a > nT$ (over-beaming). An example of inappropriate use of the divisor nT.

2.5.3 Three-Dimensional Calculation of the Total Energy Deposited in the Phantom

The integral dose E_{tot} (i.e., the total energy absorbed in the phantom) serves as a simplified indicator of patient risk: the presumption is that cancer risk increases the larger the dose and irradiation volume containing radiosensitive tissue. Denoting $f(r, z)$ as the single-rotation, axial dose profile along a given z-axis located at radius r from the central axis, and integrating over both r and z, one can calculate the total energy absorbed in the entire volume of a phantom of mass density ρ. For N adjacent rotations each spaced at interval b with respect to one another, the energy deposited in a cylindrical phantom of radius R and mass density ρ is given by the product of N and the energy deposited per single rotation,

$$E_{\text{tot}} = N\rho \int_{-\infty}^{\infty} \int_{0}^{R} f(r,z)2\pi r dr dz = \rho Nb \int_{0}^{R} 2\pi r dr \left\{ \frac{1}{b} \int_{-\infty}^{\infty} f(r,z)dz \right\} = \rho L \int_{0}^{R} D_{\text{eq}}(r)2\pi r dr \qquad (2.26)$$

where Eq. (2.7) for D_{eq} is used, and N is replaced by the spatial surrogate $L = Nb$.

Eq. (2.26) can be re-written as

$$E_{\text{tot}} = \rho \pi R^2 L \left\{ \frac{1}{\pi R^2} \int_{0}^{R} D_{\text{eq}}(r)2\pi r dr \right\} = \rho \pi R^2 L \bar{D}_{\text{eq}}, \qquad (2.27)$$

which is expressed in terms of the planar average (denoted $\bar{D}_{\text{eq}}$) of $D_{\text{eq}}(r)$ over the area πR^2 of the central scan plane, located at the midpoint $z = 0$ of the longitudinal scanning range of length L.

The following points elucidate important physics aspects of Eqs (2.26) and (2.27):

- $E_{\text{tot}} = \rho \pi R^2 L \bar{D}_{\text{eq}}$ is *not* equal to the energy deposited inside the scanned volume $\pi R^2 L$, but rather it includes significant energy deposited beyond $(-L/2, L/2)$ by scattered radiation. Therefore, division of E_{tot} by the directly irradiated mass $\rho \pi R^2 L$ does *not* equal the average dose over the scanned volume $\pi R^2 L$.

- Thus, it follows that $\bar{D}_{eq}$ is *not* equal to the average dose over the scanned volume $\pi R^2 L$.

- The relation $E_{tot} = \rho \pi R^2 L \bar{D}_{eq}$ is valid for any scanning length L, *even for sub-equilibrium scanning lengths*. So if, for a particular value of L, dose equilibrium *has not* been obtained, then to evaluate E_{tot} accurately, one must nevertheless use the equilibrium dose $\bar{D}_{eq}$ in Eq. (2.27).

- $N = t_0/\tau$ corresponds to the total x-ray beam-on time t_0, which is related to the *total mAs* on which E_{tot} fundamentally depends.

- For a given tube current and a given beam-on time t_0, i.e., for one particular *total mAs*, E_{tot} is independent of the scanning length L, where L depends on table velocity υ.

As the scanning length L increases, the cumulative dose radial distribution $D_L(z=0; r)$ becomes relatively more uniform across r due to greater scatter buildup on the central phantom axis than on the peripheral axes, and thus the equilibrium dose $D_{eq}(r)$ exhibits a weaker radial variation than the cumulative dose $D_L(z=0; r)$ for $L < L_{eq}$. Since dose measurements are typically made at only two values of r – on the phantom central ($r=0$) and peripheral ($r=R-10$ mm) axes – a "two-point" approximation to the radial integral in Eq. (2.27) can be made by assuming a plausible relative functional variation of $D_{eq}(r)$ with r. Better approximations can be made determining a more exact functional form of $D_{eq}(r)$ which could be established by measurement or Monte Carlo simulation for the particular phantom and scanner being utilized.

GLOSSARY

MDCT: multi-detector CT

Shift-invariance: translational invariance of all scan technique parameters along z (independent of z-coordinate)

τ: time for single 360° gantry rotation (typically $\tau = 1$ second or less)

t_0: total "beam-on" time for a complete scan series consisting of N rotations

N: $(t_0 / \tau) =$ total number of gantry rotations in a scan series (N may not be an integer for helical scanning)

υ: table velocity for helical scans

b: table advance per rotation (mm/rot), or *table index*

b: $b = \upsilon \tau$ for helical scans; $b =$ scan interval for axial scans

$L = Nb = \upsilon t_0$: scan length

nT: table advance producing a pitch of unity (or contiguous axial scans) often denoted by "$N \times T$"

fwhm: full width half maximum of a function

a: aperture: the geometric projection of the z-collimator aperture onto the AOR (by a "point" focal spot); also equal to the *fwhm* of the primary beam dose profile $f_p(z)$.

For MDCT $a > nT$ in order to keep the penumbra beyond the active detector length nT (called "over-beaming")

$p = b/nT$: conventional pitch

$p = b/a$: dosimetric pitch

$\Pi(z/L)$: rectangular function of unit height and width L spanning interval $(-L/2, L/2)$

$D_L(z)$: accumulated dose distribution due to a complete series of N axial or helical rotations covering a scan length $L = Nb$

$f(z)$: single rotation (axial) dose profile acquired with the phantom held stationary consisting of primary and scatter contributions denoted by $f(z) = f_p(z) + f_s(z)$

D_{eq}: limiting accumulated dose $D_L(0)$ approached for large $L > L_{eq}$ in conventional CT

L_{eq}: scan length required for the central dose $D_L(0)$ at $z = 0$ to approach within 2% of D_{eq}

L_{eq}: = 470 mm on the central axis of the 32 cm diameter PMMA body phantom

E: the total energy absorbed in the phantom (integral dose) along and about a given z-axis

REFERENCES

AAPM 2010, Report of AAPM Task Group 111, Comprehensive methodology for the evaluation of radiation dose in x-ray computed tomography, American Association of Physicists in Medicine, College Park, MD, February (2010), http://www.aapm.org/pubs/reports/RPT_111.pdf.

Bracewell R.N., *The Fourier Transform and Its Applications*, 3rd ed., McGraw-Hill, New York, (2000).

Dixon R., Radiation dose in Computed Tomography. In *Handbook of X-ray Imaging* (ed. P. Russo) 791–804. CRC Press, (2018).

Dixon R., A new look at CT dose measurement: Beyond CTDI. *Med Phys* 30, 1272–1280, (2003).

Dixon R., and Ballard A., Experimental validation of a versatile system of CT dosimetry using a conventional ion chamber: Beyond CTDI100. *Med Phys* 34(8), 3399–3413, (2007).

Dixon R., and Boone J., Dose equations for tube current modulation in CT scanning and the interpretation of the associated CTDI$_{vol}$. *Med Phys* 40, 111920 (14pp), (2013).

Dixon R., Boone J., and Kraft R., Dose equations for shift-variant CT acquisition modes using variable pitch, tube current, and aperture, and the meaning of their associated CTDIvol. *Med Phys* 4, 111906, (2014). doi: 10.1118/1.4897246.

FDA, Code of Federal Regulations, Diagnostic x-ray systems and their major components, 21CFR §1020.33, Govt. Printing Office, (August 1984).

IEC, International Standard IEC 60601-2-44, *Medical Electrical Equipment — Part 2-44: Particular Requirements for the Basic Safety and Essential Performance of X-ray Equipment for Computed Tomography*, 3rd ed., International Electrotechnical Commission, Geneva, Switzerland, (2016).

Jucius R., and Kambic G., Radiation dosimetry in computed tomography (CT). *Proc SPIE* 127, 286–295, (1977).

Leitz W., Axelson B., and Szendro G., Computed tomography dose assessment: A practical approach. *Radiat Prot Dosim* 57, 377–380, (1995).

McGee P.L., and Humphreys S., Radiation dose associated with spiral computed tomography. *J Assoc Can Radiol* 45, 124–129, (1994).

McNitt-Gray M., and Cagnon C.H., Radiation dose in spiral CT: The relative effects of collimation and pitch. *Med Phys* 26, 409–414, (1999).

Shope T., Gagne R., and Johnson, G., A method for describing the doses delivered by transmission x-ray computed tomography. *Med Phys* 8, 488–495, (1981).

CHAPTER 3

Experimental Validation of a Versatile System of CT Dosimetry Using a Conventional Small Ion Chamber

3.1 INTRODUCTION

This chapter is primarily an experimental exposition and authentication of a system of CT dosimetry (Dixon 2003) utilizing a conventional (short) ion chamber by means of which one can measure the accumulated dose at the center of the scan length (or any other point) for a scan series of any arbitrary scan length L up to the total phantom length available; the principal motivation being the fact that $CTDI_{100}$ correctly predicts the dose for only one particular scan length $L = 100$ mm and underestimates the limiting equilibrium dose (or $CTDI_\infty$) which is approached for clinically relevant body scan lengths of 250 mm or more. This is due to the integration length provided by the 100 mm long pencil chamber being too short to encompass the entire axial dose profile $f(z)$ including its very long scatter tails which also extend beyond the length of the short (14 or 15 cm) standard CTDI phantom. Dixon (2003) originally proposed an alternative to circumvent this limitation by using a helical scan of length L to translate a phantom containing a short ion chamber through the CT beam plane, thereby collecting essentially the same integral as would a pencil chamber, but rather integrated over an arbitrary length $(-L/2, L/2)$. A rigorous theoretical basis for CT phantom dosimetry has also been developed (Dixon 2003; Dixon et al. 2005) and is summarized in the next section. Other investigators (Nakonechny et al. 2005; Morgan and Luhta 2004; Anderson et al. 2005) have applied the method and equations successfully and/or supplied additional related data (Mori et al. 2005) in various phantoms and applications using a variety of detectors. AAPM Report 111 (AAPM 2010) also adopted this method, as has likewise AAPM Task Group 200 (work in progress). This method using a small (0.6cc) Farmer chamber was used by a Mayo Clinic group in making direct measurements in anthropomorphic phantoms of various sizes using clinical techniques for the purpose of determining SSDE (size specific dose estimates) for AAPM Report 204 (AAPM 2011).

The primary purpose of the present chapter is to demonstrate experimentally the implementation (and versatility) of this small ion chamber method by direct measurement of the accumulated dose in the body phantom for any desired scan length L (up to the available phantom length) *including the limiting equilibrium dose* (symbolically denoted by $CTDI_\infty$), thereby establishing the magnitude of the shortfall of $CTDI_{100}$; and further discovering if any experimental pitfalls occur while providing a practical guide for implementation. Validation is provided by comparison with pencil chamber results at the appropriate L (although the straightforward measurement method gives *prima facie* validation by itself, the pencil comparison provides added assurance). Additionally, a simple and robust method for independently verifying the pencil chamber active length is described.

A second, perhaps more important, advantage of the small ion chamber is its use in *stationary phantom* CT techniques to which the CTDI-paradigm does not apply such as wide cone beam CT without table motion or stationary table perfusion studies. If the wide cone beam in a single rotation irradiates the same length of phantom as a helical scan with phantom translation, then the dose distribution and the peak doses at $z = 0$ will be essentially the same. So, all one has to do is make a single measurement of the cone beam dose at $z = 0$ with no phantom translation using a small ion chamber to obtain the same dose as predicted for the helical scan using CTDI. One "point" measurement for the cone beam and you are done! No pencil chamber needed or desired. This is the method recommended by AAPM TG-111 (AAPM 2010). This is illustrated in greater detail in Chapter 5.

Measurements are made using both the small ion chamber method and the pencil chamber method (using pencil chambers of both 100 mm and 150 mm length) in order to compare the results of the two methodologies at the short scan lengths corresponding to these pencil lengths. Ion chamber dosimetry alone is utilized with special attention being paid to precision and accuracy, using both a 400 mm long 32 cm PMMA (Acrylic) body phantom intended to allow sufficient scan length to achieve equilibrium and the more common 150 mm long phantom for comparison.

3.2 SUMMARY OF CT DOSE THEORY IN A CYLINDRICAL PHANTOM

A brief summary of the theory (Dixon 2003; Dixon et al. 2005) pertinent to these measurements is covered in detail in Chapter 2. Parameters utilized are defined in the glossary for quick reference.

As shown in Chapter 2, the accumulated dose distribution resulting from a *helical scan acquisition* of scan length $L = \upsilon t_0$ (t_0 = total beam-on time) with a table advance per rotation of $b = \upsilon\tau$, is given by the convolution,

$$D_L(z) = \frac{1}{b} f(z) \otimes \Pi(z / L) \tag{3.1}$$

where $f(z)$ is the dose profile generated in a stationary, cylindrical phantom using a single axial rotation. On the central phantom axis (AOR) the dose rate is constant, and the actual absolute dose on the AOR is a smooth (non-oscillatory) function of z described by $D_L(z)$ in Eq. (3.1), and having a central ($z=0$) value of,

$$D_L(0) = \frac{1}{b} \int_{-L/2}^{L/2} f(z')dz' \tag{3.2}$$

For helical scans on the peripheral phantom axes where the dose distribution along z is periodic with fundamental period b, the smooth function in Eq. (3.1) and the central dose in Eq. (3.2) represent the *angular average* of the accumulated dose (Dixon 2003), averaged over all *peripheral axes at a fixed value of z* (e.g., $z=0$). The dose at the center ($z=0$) of the scan length L, as derived in Chapter 2, approaches a limiting, equilibrium value D_{eq} when L is large enough to encompass the complete scatter tails of $f(z)$, symbolically noted in Eq. (3.2) by the replacement $L \to \infty$, $D_L(0) \to D_{eq}$, and corresponding to $CTDI_\infty = (b/nT)D_{eq}$ where $p = b/nT$ is the helical pitch.

For an axial scan series, Eq. (3.1) was shown (Dixon 2003) and in Chapter 2 to represent the "running mean"; a longitudinal average over one period $b(-b/2 \le z \le b/2)$ where b is the scan interval.

3.3 ACCUMULATED DOSE (OR CTDI) MEASUREMENTS

The usual $CTDI_L$ measurement represents the accumulated dose at the center of the scan length L, $D_L(0)$ in Eq. (3.2), where $CTDI_L = (b/nT) D_L(0)$ and where nT is defined as the total reconstructed slice width acquired in a single rotation (sometimes referred to as "N × T" for MDCT). CTDI was originally defined (Shope et al. 1981) and likewise defined in the original FDA (1984) regulatory proposal with (symbolically) infinite integration limits which were expressly intended to signify that CTDI represented the limiting equilibrium dose D_{eq} for a series of contiguous scans ($b = nT$), *and that arguably remains the ideal measurement goal, since it represents an asymptotic upper dose limit which is closely approached for clinically relevant body scan lengths.* Fortunately, this measurement goal is not too difficult to attain (Figure 3.2).

3.3.1 Pencil Chamber Acquisition Method

Using a pencil chamber of active length ℓ, a single axial rotation is made about the center of the pencil chamber ($z=0$) with the phantom and table held stationary, thereby measuring the integral of the single rotation axial dose profile $f(z)$ over $(-\ell/2, \ell/2)$. This integral measurement $\int_{-\ell/2}^{\ell/2} f(z')dz'$ is not a measurement of the dose delivered by the acquisition scan (Dixon 2006), even if the integral is divided by nT to give units of dose, but can be used to *foretell* the accumulated dose $D_\ell(0)$ which would accrue in a hypothetical scan series using any value of scan interval b (or pitch b/nT) for which one wishes a dose prediction covering a phantom length L equal to the ion chamber length ($L = \ell$); and obtained by dividing the

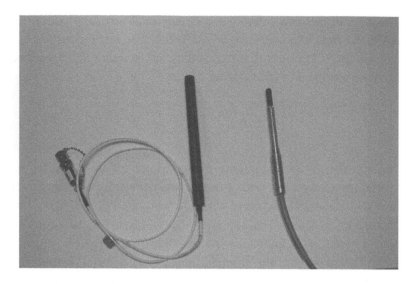

FIGURE 3.1   Picture of 100 mm long pencil chamber and 0.6cc 2571 Farmer chamber used herein.

measured integral by b, this divisor being independent of the acquisition measurement. If the divisor $b = nT$ is chosen, the predicted dose is called $CTDI_\ell$ which is equal to the dose accrued for contiguous axial scans (or helical for a pitch = 1) covering a scan length ℓ.

This method is quite restrictive, since a 100 mm long pencil chamber can only accurately predict the accumulated dose at $z = 0$ for a scan length of exactly 100 mm, and underestimates the integral in Eq. (3.2) and thence $D_L(0)$ for longer scan lengths $L > \ell$ (as shown in Figure 3.2).

3.3.2 Small Ion Chamber Acquisition Method

With the small ion chamber method (Dixon 2003; Nakonechny et al. 2005; Morgan and Luhta 2004; Anderson et al. 2005; Dixon et al. 2005), the accumulated dose at $z = 0$ [$D_L(0)$ in Eq. (3.2)] is directly measured (rather than predicted) by integrating the current from an ion chamber located at a fixed point in the phantom at the midpoint ($z = 0$) of the scanned length $L = \upsilon t_0$, while the phantom and ion chamber are translated by the couch through the beam plane at velocity υ during a helical acquisition of total (x-ray on) time t_0. The charge collected by the ion chamber q_h(nC) is converted to accumulated dose as $D_L(0) = N_k q_h$, where N_k is the chamber calibration factor in mGy/nC. If $D_L(0)$ is multiplied by the acquisition pitch $p = b/nT$, it can be converted to $CTDI_L$. The acquisition pitch p can be arbitrary (Dixon et al. 2005), since the dose delivered for any other desired (clinical) pitch $\tilde{p}$ can subsequently be obtained by multiplying the measured dose $D_L(0)$ by the pitch ratio $(p/\tilde{p}) = (b/\tilde{b})$, or computed from $CTDI_L$ using the inverse $\tilde{D}_L(0) = (\tilde{p})^{-1} CTDI_L$.

Using the small ion chamber, the measurement method itself thus guarantees that the scan length = υt_0 is *always identical* to the integration length L of the single-rotation dose profile $f(z)$ in Eq. (3.2); and it is also self-evident that one is directly measuring the accumulated dose at $z = 0$ in the phantom during a scan series of length $L = \upsilon t_0$. The utility of this method has been demonstrated by other investigators (Dixon 2003; Nakonechny et al. 2005;

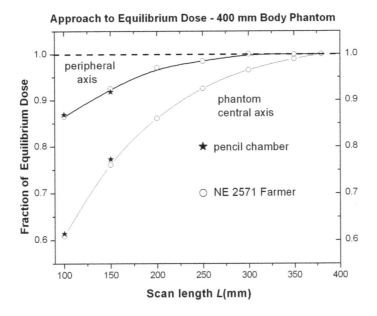

FIGURE 3.2 Approach of accumulated dose $D_L(0)$ in Eq. (3.2) to the equilibrium dose D_{eq} as scan length $L \to \infty$. Plotted is the fraction $f_L = D_L(0)/D_{eq}$ measured in the 400 mm long body phantom on a GE LS-16 scanner at 120 kVp using $nT = 16 \times 1.25$ mm $= 20$ mm.

(From Dixon and Ballard, *Medical Physics*, 2007.)

Morgan and Luhta 2004; Anderson et al. 2005). It also allows one to emulate the results that would be obtained using a "virtual pencil chamber" of any arbitrary length $\ell = L$, up to the total phantom length available, and also to calculate the $CTDI_L$. Thus, the pencil chamber, of fixed and limited length, can be replaced by this methodology which exhibits much greater flexibility and allows measurement of the accumulated dose for any desired scan length, or the CTDI for any integration length – not just 100 mm; moreover, it allows one to measure the equilibrium dose D_{eq} (or $CTDI_\infty$) if the phantom is sufficiently long to allow it. Measurement of $CTDI_\infty$ on the AOR would require a pencil chamber more than 400 mm long.

However, as with any measurement technique, certain guidelines and caveats should be observed (Dixon et al. 2005) which are clearly illustrated while describing the following experimental design and results.

3.3.3 The Measured Quantity – A Phantom Dose Surrogate

CT dose measurements using an ion chamber in a phantom "cavity" differ from such measurements in megavoltage (MV) beams, since the primary electrons originating in the phantom material have insufficient energy to penetrate the ion chamber walls and cannot contribute to the charge q collected. The ion chamber is simply measuring the exposure or air kerma existing in the cavity which is *ideally* given by qN_k (we tacitly assume a temperature–pressure correction is also made). The *actual* air kerma in the cavity is given by $D_{air} = qN_k k_Q$ where k_Q is an additional (and subtle) chamber correction factor (Ma and

Nahum 1995) owing to the fact that the chamber is calibrated in air using a mono-directional beam incident normal to its axis, whereas the exposure in the cavity may be more nearly isotropic (depending on the scatter to primarily ratio S/P); but the response of a cylindrical chamber is not entirely isotropic – exhibiting a loss for end-on photon incidence, thereby requiring a slight boost to its reading ($k_Q > 1$). There are also differences in photon spectra at various cavity locations (Mayajima 2002–18) (differing from the calibration spectrum as well) which may also affect k_Q; however, the "flat" energy response of the Farmer chamber minimizes this effect for CT spectra.

The dose to the phantom medium itself with the cavities filled is given by $D_{med} = qN_k k_Q p_{dis}(\bar{\mu}_{en}/\rho)_{air}^{med}$ where p_{dis} is the *displacement factor* utilized to correct for the phantom material displaced by the cavity, the missing matter producing both a loss of attenuation ($p_{dis} < 1$) and a compensatory loss of adjacent scatter ($p_{dis} > 1$) the net value depending on photon energy. For a PMMA phantom having poor tissue (or air) equivalence, it is desirable to use units of air kerma, thus we can simply imagine that the cavities are filled with a solid air-equivalent material having the same density as the phantom, in which case the last term $(\bar{\mu}_{en}/\rho)_{air}^{med} = 1$.

These correction factors have been investigated in considerable detail for the NE 2571 Farmer chamber using Monte Carlo calculations (Ma and Nahum 1995; Seuntjens and Verhaegen 1996) [but only for mono-directional kVp therapy beams in water over a limited range of depths (2–5 cm) and field sizes (up to 200 cm²)] and show that the *product* $k_Q p_{dis}$ is typically less than 1.02 at beam qualities comparable to CT (Ma and Nahum 1995). In our case, the S/P ratio is likely much larger due to field sizes of up to 1,000 cm² and depths up to 16 cm for the AOR, plus the CT beam geometry is rotational; hence it would be speculative to extrapolate these values of $k_Q p_{dis}$ to CT measurement using the same chamber. Nevertheless, it seems unlikely that $k_Q p_{dis}$ will exceed 1.02 – at least for the Farmer chamber. There is greater uncertainty in the case of pencil chamber with its anomalous elongated cavity.

So as not to be unduly distracted from our goal, we will take the same simplified approach used with $CTDI_{100}$, and assume that qN_k is equal to the actual air kerma in the phantom cavity (ignoring k_Q) and also assume that it is also representative of the dose to the phantom at the cavity location (but expressed in units of air kerma) – also ignoring the cavity displacement factor p_{dis}.

3.4 MATERIALS AND METHODS

A 400 mm long, 32 cm diameter cylinder, Acrylic (PMMA) dosimetry body phantom was used having measurement holes drilled along its central axis and peripheral axes 1 cm below its surface.

All the ion chambers utilized, and their properties are listed in Table 3.1, each having received a recent (2006) accredited dosimetry calibration laboratory (ADCL) calibration. The same calibrated Keithley model 616 electrometer with electronic bias supply set to 300 v. was used with all ion chambers.

In order to create a snug cavity for the NE 2571 Farmer chamber in the CT dosimetry phantoms, the Acrylic Co-60 buildup cap was reduced to 1.25 cm diameter in order to fit

TABLE 3.1   Ion Chambers and Properties

Make/Model	Volume (cm³)	Active Length ℓ (mm)
Victoreen 500–200	10 cc	100 mm
Capintec PC-4P14	4.5 cc	152 mm
Nuclear Enterprise 2571 (Farmer type)	0.6 cc	≈23 mm

Source: Dixon and Ballard, *Medical Physics* (2007).

in the cylindrical phantom holes; the cable inside the phantom was fitted with a Delrin sleeve; and a PMMA (Acrylic) rod inserted in the opposite end to fill the remaining void.

Representative dose data on the peripheral axis vs. *z* for helical scan series were obtained using 150 mm long OSL (Al$_2$O$_3$) ribbons (Peakheart et al. 2003) supplied and read out by Landauer, Inc. (Landauer, Inc. Glenwood, IL).

3.5 MEASUREMENTS VALIDATING THE PRECISION AND ACCURACY OF THE DOSIMETRY SYSTEM ITSELF

3.5.1 Test of the "Stem Effect" for NE 2571 Farmer Chamber

With the small ion chamber method, for long scan lengths one is irradiating the entire Farmer chamber stem as well as part of its cable. A test made using two irradiations with perpendicular chamber orientations on the central axis of an elongated 6×46 cm field (free-in-air using a heavily filtered, 120 kVp, diagnostic x-ray beam) *produced less than 0.4% change* in response despite irradiating more than 200 mm of stem and cable.

Reversal of the bias polarity had no detectable effect on the chamber reading.

3.5.2 Effect of Phantom Cavity on Farmer Chamber Reading

Use of the "buildup cap," cable sleeve, or opposing rod had little effect on the reading. Inserting the bare Farmer (without cap, sleeve, or opposing rod), gave an increased reading of only +0.6% on the peripheral axis, and +0.9% on the central phantom axis. Removing both the opposing Acrylic rod and cable sleeve resulted in only a 0.2% increase in the "cap on" Farmer reading.

3.5.3 Cross Comparison of Ion Chambers

Since the absolute ADCL chamber calibration factors have a stated 2σ (95% confidence) uncertainty of $\pm 5\%$ for the pencil chambers and $\pm 1.5\%$ for the NE 2571 Farmer chamber, the credibility and precision of our data comparing pencil vs. Farmer chambers can be enhanced by a direct cross-comparison of the chambers in the actual CT beam utilized. This is done by making a free-in-air measurement of the charge integrated on the AOR during a helical scan, whereby the ion chambers are completely translated through the primary beam profile $f_p(z)$, while attached to the couch but extending beyond its end in order to provide a relatively scatter-free environment. It is straightforward to show that the *integrated charge q_h measured during this helical acquisition* is given by,

$$N_k q_h = \frac{1}{b} \int_{-\infty}^{\infty} f_p(z')dz' \tag{3.3}$$

where $b = \upsilon\tau$ is the table advance per rotation. For the same acquisition pitch $p = b/T$, the chamber readings $N_k q_h$ (mGy) should all be the same (independent of chamber length), assuming the calibration factors N_k are accurate.

Using this method, the chamber charge response q_h depends only on N_k^{-1} and is independent of the active length ℓ of the ion chamber, since each incremental element of chamber length samples the entire primary beam profile $f_p(z)$.

Table 3.2 shows the measured ion chamber responses ($N_k q_h$) normalized to that for the NE 2571 Farmer chamber.

The relative response shown in the first column indicates that the ADCL calibration factors appear to have considerably greater accuracy than quoted – at least in this case. For further validation, values previously measured in 2003 (Dixon 2003), using the identical ion chambers and method but N_k values from the previous calibration, are nearly identical, as shown by the second column data.

The chamber active length ℓ introduces additional measurement uncertainty *for the pencil chamber methodology*. The charge q collected by the pencil chamber in-phantom during the single axial rotation about its center *does not represent a dose* (nor does qN_k), but rather a dose line integral having units of mGy·cm; therefore knowledge of N_k alone is insufficient and one must also utilize the active length ℓ supplied by the manufacturer in order to compute the average chamber response per unit length, $\bar{c}_\ell = (N_k \ell)^{-1}$ in units of nC·cm^{-1}mGy^{-1}. Calibration laboratories in the US typically measure only N_k (mGy/nC) by uniformly irradiating the entire chamber length, and any additional "per unit length" factors provided in the report are *calculated* values based on the manufacturer-specified value of ℓ rather than measured values.

If the pencil chamber response per unit length is denoted by the function $c(z)$ (nC·cm^{-1}mGy^{-1}), then the *charge q_{ss} collected in a single axial rotation* is,

$$q_{ss} = \int\limits_{-\ell/2}^{\ell/2} c(z) f(z) dz \tag{3.4}$$

In order to faithfully reproduce the integral of $f(z)$, which is the measurement goal [see Eq. (3.2)], we require that $c(z)$ be uniform over the entire chamber length, ideally $c(z) = c(0)\Pi(z / \ell)$; however, this is physically impossible (even if the chamber has perfect

TABLE 3.2   Ion Chamber Cross-Comparison in Air, Normalized to the NE 2571 Farmer Chamber Response, Using the Actual CT Beam and a Helical Acquisition at 120 kVp and the ADCL-supplied Calibration Factors N_k for Each Chamber

Ion Chamber	Relative Response (2006)	Relative Response (2003)	Axial vs. Helical Response ratio $[\ell q_{ss}/bq_h]$
NE 2571 (0.6cc) Farmer	1.000	1.000	N/A
Capintec PC-4P14 150 mm pencil	1.003	1.001	0.983
Victoreen 500–200 100 mm pencil	1.003	0.990	0.986

Source: Dixon and Ballard, *Medical Physics* (2007).

symmetry), since $c(z)$ must approach zero at the ends of the collection volume, and cannot do so abruptly. That is, the angular spread of the Compton electron shower (charge LSF) produced by a knife-edge primary beam impulse determines the distance from the end over which $c(z)$ begins its fall to zero.

It is argued (Bochud et al. 2001; IEC 1997) that the most important values of $c(z)$ are those near the center where $f(z)$ is largest, viz. $c(0)$; hence use of its average value $\bar{c}_\ell = (N_k\ell)^{-1}$ may underestimate $c(0)$. This is more important for "free-in-air" pencil chamber measurements, and this argument is weakened somewhat for in-phantom measurements in which $f(z)$ may still be appreciable near the ends of the chamber. Nevertheless, if ℓ is chosen properly as the *fwhm* of the chamber response function $c(z)$, then the problem resolves itself with the result that, $\bar{c}_\ell = (N_k\ell)^{-1} \cong c(0)$.

3.5.4 Validation of the Manufacturer-Supplied Pencil Chamber Active Length ℓ

Direct measurement of $c(z)$ using a slit beam is more complex than one might imagine, requiring a correction due to slit scatter (Bochud et al. 2001; IEC 1997; Jensen et al. 2006), and a *much simpler validation* is described in this section. We have already derived a simple and robust test for determining the adequacy of the value of ℓ supplied with the pencil chamber; namely, comparing the results of the free-in-air measurements on a given chamber using both the helical method of Eq. (3.3) and the single axial rotation method of Eq. (3.4). It is straightforward to show (using these two equations) that if the ratio $(\ell q_{ss}/bq_h)$ for a given pencil chamber is equal to unity, then $\bar{c}_\ell = (N_k\ell)^{-1} \cong c(0)$. Moreover, the ratio $(\ell q_{ss}/bq_h)$ is approximately independent of the extraneous ambient scatter present during the free-in-air measurements, and therefore a robust test of ℓ.

This ratio, measured as described, together with the manufacturer-supplied values of ℓ given in Table 3.1, is shown in Table 3.2 to be close to 0.985 for both chambers, thus indicating that the stated values of ℓ exhibit adequate accuracy for our purposes

3.6 BODY PHANTOM MEASUREMENTS AND RESULTS

Having verified the precision and accuracy of our dosimetry systems as described in the previous section, we proceed with the measurements in the body dosimetry phantom.

Since phantoms of two different lengths h are being used ($h = 150$ mm and 400 mm), the notation $_hCTDI_L$ will be used to denote CTDI measured in a 32 cm diameter PMMA body phantom of length h for an integration (or scan) length L. Two models of GE MDCT scanners, viz., a GE Lightspeed "16-slice" and a VCT "64-slice" (both 100 kW), were used for all measurements, which were all made using 120 kVp, a gantry rotation time $\tau = 1$ sec, the "bow- tie" filter associated with the largest body FOV, and with a 32 cm diameter PMMA "body" phantom.

3.6.1 Effect of Phantom Length on $CTDI_{100}$

The specified (FDA 1984) body dosimetry phantom is too short ($h = 140$ mm) to allow establishment of scatter equilibrium at its center, resulting in an underestimate even of $CTDI_{100}$ compared to that measured in a phantom of realistic length. That is, adding additional length to the phantom ends backscatters enough additional photons to measurably affect the integral in Eq. (3.2) over (−50 mm, 50 mm).

This effect can be measured with high precision, simply by centering the same 100 mm long pencil chamber in both the 150 mm and 400 mm long phantoms and taking the ratio of the two chamber ionization readings (q) for the same scan technique.

(A 140 mm long phantom was not available). Table 3.3 shows this long/short phantom $CTDI_{100}$ ratio measured on a GE LS-16 CT scanner.

The observed percent increase is significant (+7.2%) for the central phantom axis, but only 1.3% for the peripheral axes due to the smaller S/P ratio there; however, the percent increase may be somewhat larger when compared to the standard 140 mm long phantom (Mori et al. 2005).

3.6.2 Experimental Plan for Demonstration of the Small Ion Chamber Acquisition Method

The versatility and extended capability of this method is illustrated using the long (400 mm) phantom and NE 2571 Farmer chamber as follows.

Observe the approach to equilibrium $[D_L(0) \to D_{eq}]$ by measuring $D_L(0)$ as L is systematically increased from $L = 100$ mm to 380 mm; from which one can compute $CTDI_L = p \times D_L(0)$, where $p = b/nT$ is the arbitrary acquisition pitch utilized [see Eq. (3.2)], including (within 96%) of the equilibrium dose D_{eq} and $CTDI_\infty = pD_{eq}$. At the time of these measurements, it was thought that 380 mm was long enough to achieve equilibrium; however, later work showed that at 380 mm the central dose is actually short of D_{eq} on the body phantom central axis by 4%, and 470 mm is required to get within 2% of D_{eq}. The approach is asymptotic, so what is referred to as D_{eq} in the following may actually be 0.96 D_{eq}.

Compare dose values obtained using the Farmer to those using the pencil chambers. The pencil chamber of length ℓ is limited to predicting the accumulated dose for only one particular scan length $L = \ell$, whereas the small ion chamber can measure the dose for any arbitrary scan length L up to the total length of phantom available. For the particular scan lengths $L = 100$ mm and $L = 150$ mm, we compare $_{400}CTDI_L$ obtained using the Farmer chamber to the values $_{400}CTDI_\ell$ *in the same long phantom* using the two pencil chambers having corresponding nominal lengths $\ell = 100$ mm and $\ell = 150$ mm; thereby testing the validity of the pencil chamber methodology against that of the Farmer chamber methodology.

TABLE 3.3   Measured $CTDI_{100}$ Ratio Using the Same 100 mm Pencil Chamber in the Long/Short (400 mm/150 mm) Body Phantoms (GE LS-16, 120 kVp, 600 mA × 1 sec)

Collimator Configuration		$_{400}CTDI_{100}/_{150}CTDI_{100}$	
$n \times \Delta T$	nT (mm)	Central Phantom Axis	Peripheral Phantom Axis
16 × 1.25 mm	20 mm	1.072	1.012
8 × 1.25 mm	10 mm	1.073	1.014

Source: Dixon and Ballard, *Medical Physics* (2007).

3.6.3 Selection of Scan Parameters for the Small Ion Chamber Acquisition

The accumulated dose on the central phantom axis (AOR) is always smooth, and any value of the acquisition pitch can be used, and $p \approx 1$ is reasonable.

However, on the peripheral axes the longitudinal dose variation along z is oscillatory of period b, thus when using a *small ion chamber of active length* ℓ, one should ensure that $b < \ell$ (by selecting an acquisition pitch $p < \ell/nT$), such that the chamber will "average out" the oscillations over its length ℓ). There is no compromise associated with such averaging, since both $CTDI_L$ and $D_L(0)$, by definition, represent an average dose on the peripheral axis for both helical and axial scans (the pencil chamber also *predicts* such an average dose). For the Farmer chamber, $\ell \approx 23$ mm, hence for $nT = 20$ mm one could choose a pitch $p \le 1$ for the peripheral axis measurement; however, a pitch of unity will leave small gaps on the surface of the body phantom (Dixon 2003; Dixon et al. 2005), and a smaller pitch ($p < 0.75$) is preferable, not only producing a smaller amplitude of oscillation (less than $\pm 10\%$) but additionally improving the averaging provided by the chamber length ℓ. For example, choosing a pitch $p = 0.5$ when using $nT = 20$ mm, such that $b = 10$ mm, results in $\ell > 2b$, allowing the chamber length to average over two periods of oscillation.

3.6.4 Results of Measurements Made in the 400 mm Long Body Phantom

Measurements of $D_L(0)$ were made on a GE Lightspeed 16-slice scanner for $nT = 16 \times 1.25$ mm = 20 mm at 120 kVp, for a variety of scan lengths $L = \upsilon t_0$, including $L = 100$ mm and $L = 150$ mm, in order to compare with the measurements made using the two pencil chambers of the same nominal lengths; using a pitch $p = 0.938$ for the AOR, and $p = 0.563$ for the peripheral axis. The results, expressed as a fraction of the equilibrium dose $f_L = [D_L(0)/D_{eq}]$ as a function of L are shown in Figure 3.2.

The same data shown in Figure 3.2 are given in Table 3.4. as absolute dose values per 100 mAs, having been first converted to $CTDI_L = p \times D_L(0)$ where p is the selected *acquisition* pitch as given above. Once converted to $CTDI_L$, the acquisition pitch becomes irrelevant, and one can find the dose for any desired value of clinical pitch p_c using the inverse $p_c^{-1} CTDI_L$.

A similar set of measurements shown in Table 3.5 was performed on a GE VCT 64-slice scanner at 120 kVp, for nominal beam widths of $nT = 64 \times 0.625$ mm = 40 mm, and $nT = 32 \times 0.625 = 20$ mm, for scan lengths of $L = 100$ mm, 150 mm, and 380 mm.

The Farmer chamber data in Tables 3.4 and 3.5 for both the LS-16 and VCT scanners, normalized to the same equilibrium doses and corrected to the same collimator aperture setting, are plotted in Figure 3.3 to more closely examine the approach-to-equilibrium functions. The curves are found to be nearly coincident for the central axis, and only diverge slightly at $L = 100$ mm on the peripheral axis, as shown in Figure 3.3; however, this divergence cannot be attributed entirely to experimental error, since the 100 mm pencil chamber backup measurements agree closely with the Farmer (to within 0.5% at $nT = 20$ mm – see Table 3.7).

The average data have been fit using an exponential growth function, shown as the continuous curve in Figure 3.3 with the fit function also shown. The approach to equilibrium

TABLE 3.4   Measured $_{400}CTDI_L$ Values (mGy/100 mAs) for the GE Light Speed-16

Scan Length	Detector Width	Body-Central Axis $_{400}$CDTI$_L$ (mGy/100 mAs)		Body-Peripheral Axis $_{400}$CDTI$_L$ (mGy/100 mAs)	
L(mm)	nT (mm)	Farmer	Pencil	Farmer	Pencil
100 ± 1	20 mm	5.40	5.57	10.83	10.89
150 ± 1	20	6.74	6.86	11.57	11.49
300 ± 1	20	8.54		12.51	
350 ± 1	20	8.77		12.51	
380 ± 1	20	8.86		12.51	

Source: Dixon and Ballard, *Medical Physics* (2007).

Measured in a 400 mm long, 32 cm diameter PMMA body phantom using a small NE 2571 Farmer ion chamber with a helical acquisition, together with those values measured conventionally in the same phantom using the 100 mm and 150 mm pencil chambers; and using the calibration factors supplied by the ADCL for each of the three chambers. Helical scan lengths $L = \upsilon t_0$ used are within ±1 mm of the nominal value listed.

Acquired using $nT = 16 \times 1.25$ mm = 20 mm, 120 kVp, 350 mA × 1 sec, large FOV, large focal spot, with helical pitches of $p = 0.938$ for the AOR, and $p = 0.563$ for the peripheral axis.

TABLE 3.5   Measured $_{400}CTDI_L$ Values (mGy/100 mAs) for the GE VCT 64-Slice

Scan Length	Detector Width	Body-Central Axis $_{400}$CDTI$_L$ (mGy/100 mAs)		Body-Peripheral Axis $_{400}$CDTI$_L$ (mGy/100 mAs)	
L (mm)	nT (mm)	Farmer	Pencil	Farmer	Pencil
100 ± 1	20	5.70	5.73	11.53	11.58
150 ± 1	20	7.05	7.23	12.78	12.80
380 ± 1	20	9.30	–	13.93	–
100 ± 1	40	5.30	5.30	10.48	10.90
150 ± 1	40	6.58	6.73	11.80	11.55
380 ± 1	40	8.73	–	12.95	–

Source: Dixon and Ballard, *Medical Physics* (2007).

Acquired at 120 kVp, 400 mA × 1 sec, large FOV, large focal spot,

$nT = 64 \times 0.625$ mm = 40 mm ($p = 0.984$ on AOR and $p = 0.516$ on periphery)

$nT = 32 \times 0.625$ mm = 20 mm ($p = 0.969$ on AOR and $p = 0.531$ on periphery).

Measured in a 400 mm long, 32 cm diameter PMMA body phantom using a small NE 2571 Farmer chamber, together with values measured using the 100 mm and 150 mm pencil chambers, using the calibration factor supplied by ADCL for each of the three chambers. Helical scan lengths $L = \upsilon t_0$ used are within ±1 mm of the nominal value listed.

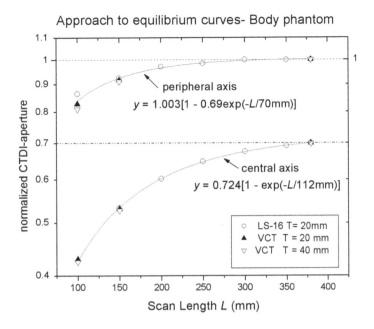

FIGURE 3.3   Approach to equilibrium re-examined 120 kVp.

(From Dixon and Ballard, *Medical Physics,* 2007.)

is determined by the length of the scatter tails and thence to photon energy or beam quality (kVp and filtration), however, the fraction $f_L = D_L(0)/D_{eq}$ is practically independent of beam width or nT.

3.7 ANALYSIS OF BODY PHANTOM DATA

3.7.1 Underestimation of Equilibrium Dose in the 400 mm Phantom Due to Truncation of Integration Length to 100 mm

It is evident from Figures 3.2 and 3.3 that a scan length of $L = 100$ mm results in an accumulated dose D_{100} (0) which is only a fraction of the equilibrium dose D_{eq}; this fraction $f_{100} = [D_{100}(0)/D_{eq}]$ in the 400 mm long phantom is shown in Table 3.6 for both the GE LS 16-slice and VCT 64-slice scanners; together with the directly comparable fractions reported by Boone (2007) and Mori et al. (2005) in a PMMA phantom of the same diameter. The Boone (2007) data are from a Monte Carlo simulation in an infinitely long phantom, and that of Mori et al. (2005) in a like phantom 900 mm long.

The simulations of Boone (2007) indicated that the ratio f_{100} as tabulated above is approximately independent of nT, decreasing by only about 1% in going from very narrow nominal beam widths $nT = 1$ mm to $nT = 40$ mm (see Appendix A for rationale).

The central axis value we obtained $f_{100} = 0.62$ seems quite solid, independent of nT (at least over 20–40 mm) and agreeing well with both Mori et al. (2005) and Boone (2007); and the average of our peripheral axis values $\bar{f}_{100} = 0.83$ seems a reasonable value to choose as a basis for further discussion, being close to the global average of 0.84.

But the factors given in Table 3.6 do not represent the totality of the shortfall of $CTDI_{100}$.

TABLE 3.6 Fraction (f_{100}) of the Equilibrium Dose D_{eq} Attained for a Scan Length $L = 100$ mm, in a 32 cm diameter PMMA body phantom having length $h \geq L_{eq}$; where $f_{100} = [D_{100}(0)/D_{eq}] = [CTDI_{100}/CTDI_{\infty}]$

Investigator	Scanner	Method	Phantom Length	nT (mm)	$D_{100}(0)/D_{eq}$ Central Axis	$D_{100}(0)/D_{eq}$ Peripheral Axis
This work	GE LS-16	Farmer	400 mm	20 mm	0.62	0.86
	GE VCT-64	Farmer	400 mm	20 mm	0.62	0.83
				40 mm	0.62	0.81
(Boone 2007)	GE model	Monte Carlo	Infinite	20 mm	0.63	0.88
				40 mm	0.62	0.87
(Mori et al. 2005)	Toshiba-256	Diode	900 mm	20 mm	0.59	0.79
Global average				Average	<0.62>	<0.84>

Source: Dixon and Ballard, *Medical Physics* (2007).

3.7.2 Underestimation of Equilibrium Dose by $CTDI_{100}$ – Total Shortfall

One is tempted to say that $CTDI_{100}$ underestimates the equilibrium dose by the ratio $f_{100} = {}_{400}CTDI_{100}/{}_{400}CTDI_{\infty}$ as obtained from Table 3.6, viz. 0.62 for the central axis and 0.83 for the peripheral axes; however, $CTDI_{100}$ is not typically measured in a phantom of such realistic length, but rather in one of 140 or 150 mm length, hence we must also divide by the additional factors shown in Table 3.3, with the result that:

The value of $_{150}CTDI_{100}$ as currently determined, underestimates the limiting equilibrium dose in the 32 cm diameter PMMA body phantom by at least a factor of 0.58 on the central axis, and by 0.82 on the peripheral axes. Thus, one should multiply $_{150}CTDI_{100}$ by the reciprocal factors of 1.72 on the AOR, and 1.22 on the peripheral axes in order to obtain the equilibrium dose (or $CTDI_{\infty}$) – at least for the particular GE scanners tested herein at 120 kVp.

3.7.3 Farmer vs. Pencil Chamber Comparison

A comparison of $CTDI_L$ using the small ion chamber method vs. the pencil chamber method in the case where $\ell = L$.

As previously illustrated in Figure 3.2, and shown numerically in Tables 3.4 and 3.5, the $_{400}CTDI_L$ values for $L = 100$ and 150 mm determined using the pencil and Farmer chambers show remarkable agreement considering the disparity in chamber type and cavity geometry, as well as a differing acquisition method (which relies on specification of the active length parameter ℓ in addition to N_k for the pencil chambers). This agreement is summarized in Table 3.7, expressed as the *ratio* of pencil to Farmer chamber dose values from the measurements made in the same 400 mm phantom on both the LS-16 and VCT scanners.

Agreement is quite good, with an overall (global) average within 1% for both pencil chamber lengths; with no real systematics observed beyond the expected experimental error of a ratio. Nakonechny et al. (2005) likewise obtained good agreement between a 100 mm pencil and a small ion chamber (Wellhöfer IC-10).

As mentioned, this agreement serves more as a validation of the pencil chamber methodology than that of the small ion chamber; and *the fact that one can indeed measure $CTDI_{100}$*

TABLE 3.7   Farmer vs. Pencil Chamber Comparison

| | | Ratio of Measured $_{400}CTDI_L$ Values (Pencil to Farmer) | | | |
| | | L = 100 mm | | L = 150 mm | |
Scanner	nT (mm)	Central Axis	Peripheral Axis	Central Axis	Peripheral Axis
GE LS-16	20 mm	1.030	1.005	1.017	0.993
GE VCT-64	20 mm	1.004	1.004	1.025	1.002
	40 mm	1.000	1.040	1.022	0.978
Average ± std. dev.		1.011 ± .015	1.016 ± .020	1.021 ± .004	0.991 ± .012
Global average		1.01 ± .013			

Source: Dixon and Ballard, *Medical Physics* (2007).

using the small ion chamber method is not of primary interest, since a scan length of 100 mm is not representative of clinical body scans lengths which may exceed 250 mm or more. Since we now have the capability of measuring $D_L(0)$ or $CTDI_L$ for any scan length L, we should rather strive to measure the limiting equilibrium value $CTDI_\infty$ from which the equilibrium dose D_{eq} can be computed for any desired and arbitrary value of generalized pitch $p = b/nT$ from $D_{eq} = p^{-1} CTDI_\infty$. That notwithstanding, the value of $CTDI_{100}$ may still have interest as a reference for historical comparison, dose normalization, or as a bridge to other dosimetry data.

The $CTDI_\infty$ values measured for the GE LS-16 scanner (from Table 3.4) are 8.9 mGy/100 mAs on the central axis (AOR), and 12.5 mGy/100 mAs on the peripheral axes.

3.8 CTDI-APERTURE

3.8.1 A Useful Constant Deriving from Conservation of Energy and a Robust Measurement Shortcut

It has been shown (Dixon et al. 2005) that the *infinite* integral of $f(z)$ is proportional to the collimator aperture in the z-direction, denoted by its point-projected value a on the AOR, which leads to the constancy of "CTDI-aperture" which we defined (Dixon et al. 2005) as,

$$CTDI_a = \frac{1}{a} \int_{-\infty}^{\infty} f(z)dz = \text{constant} \tag{3.5}$$

Its constancy is even more fundamental than the general model from which it was deduced (Dixon et al. 2005), and results directly from the conservation of energy, since the total energy incident on the phantom is proportional to the z-aperture a. For aperture settings large compared to the penumbra, the primary beam *fwhm* is equal to a; however, the *fwhm* has no physical significance relative to the constancy of $CTDI_a$ which remains constant even in the region where a is small enough (Dixon et al. 2005) such that the *fwhm* ≠ a.

For the GE MDCT scanners, the aperture setting is electro-mechanically measured, with its value tightly controlled about a nominal value by a closed loop tracking system (Toth et al. 2000), and values are verified daily by the system on startup.

Thus, for these scanners one needs only to measure the dose for one known value of the aperture in order to compute $CTDI_a$, from which the entire table of CTDI values can be generated using

$$CTDI_\infty = (a/nT)CTDI_a \qquad (3.6)$$

For MDCT, the factor $(a/nT) > 1$ is the "over-beaming" factor, since $a > nT$ is required in order to extend the penumbra beyond the active detector length nT.

Table 3.8 lists the typical aperture values for the GE VCT which can be obtained from the dose efficiencies $100(a/nT)^{-1}$ displayed on the scanner monitor for the various available acquisition configurations $nT = n \times 0.625$ mm. Although these aperture values are typical values, their variation from scanner to scanner is estimated (Toth et al. 2000) to be within 1% for the larger beam widths ($nT \geq 10$ mm), and within 2.5% for $nT \leq 5$ mm.

Note that CTDI (and $CTDI_{air}$ measured free-in-air [Dixon et al. 2005]) linearly track the over-beaming factor a/nT in Table 3.8 and thus vary significantly with nT, while $CTDI_a$ remains constant.

Using the aperture values in Table 3.8, and the measured $CTDI_\infty$ from Table 3.6, it is easy to verify that the computed $CTDI_a$ at $nT = 20$ mm and $nT = 40$ mm agree within 0.1% on the AOR, and within 1% on the peripheral axis (independent of nT as predicted).

Since the definition of $CTDI_a$ involves the infinite integral of $f(z)$, it rigorously applies only to the equilibrium dose $CTDI_\infty$ [Eq. (3.6)]; however, since the ratio of $CTDI_{100}/CTDI_\infty$ was shown by Boone (2007) to be nearly constant, varying by only 1% over (1 mm $<$ nT $<$ 40 mm), for all practical purposes $CTDI_{100}$-aperture is also constant. As corroboration, the values of $(CTDI_a)_{100}$ for the GE LS-16, derived from the measured $CTDI_{100}$ data in the technical manual and the published (Dixon et al. 2005) typical aperture values, illustrate its constancy to within a few percent on both the central and peripheral axes in the head and body phantoms for both large and small focal spot sizes, as shown in Table A.1 (Appendix A).

TABLE 3.8   GE VCT Scanner Aperture Settings

Acquisition Collimator		Focal Spot Size			
		Large		Small	
$n \times \Delta T$	nT (mm)	a (mm)	a/nT	a (mm)	a/nT
64 × .625	40	42.13	1.053	41.84	1.046
32 × .625	20	22.39	1.12	21.57	1.08
16 × .625	10	13.11	1.31	12.13	1.21
8 × .625	5	8.229	1.65	7.049	1.41
4 × .625	2.5	3.705	1.48	3.771	1.51
2 × .625	1.25	2.459	1.97	1.836	1.47

Source: Dixon and Ballard, *Medical Physics* (2007).
a = collimator aperture (point) projected onto the AOR.
a/nT = over-beaming factor.
$100(a/nT)^{-1}$ = dose efficiency % displayed on CT console.

Given the apertures for the GE MDCT scanners in Table A.1 (or Table 3.8), only four CTDI measurements (on the AOR and peripheral axis in the head and body phantoms) will, in theory, allow one to completely fill in all 64 CTDI values in Table A.1 (or all 24 values for the $n \times 0.625$ mm modes for the VCT), as was specifically verified and shown in Table A.1 (Appendix A), and also confirmed by our own data over a more limited range of apertures.

3.9 VALIDATION OF PERIPHERAL AXIS DATA

3.9.1 Correcting a Misconception

A question which has arisen frequently enough to warrant the additional explanation and verification in this section concerns the accuracy of the small ion chamber method on the peripheral axes, considering the often large values of pitch ($p \geq 1$) used in clinical protocols, for which both the period and amplitude of the dose oscillations may be large. This concern typically arises due to the misconception that the acquisition scan must necessarily emulate the clinical technique for which one desires the dose (including an identical pitch) which is not the case. Since the accumulated dose is *rigorously* (Dixon 2003; Dixon et al. 2005) proportional to 1/pitch (p^{-1}) as can be seen from Eqs (3.1 and 3.2); an arbitrarily small acquisition pitch $p = b/nT$ can be used, and the dose scaled to any other desired pitch p' using the pitch ratio p/p' as a scaling factor. Thus, the acquisition and the clinical scan techniques can be uncoupled with respect to pitch.

The dose measured in the acquisition scan at pitch p can be converted to $CTDI_L = pD_L(0)$. Once $CTDI_L$ has been calculated, the acquisition pitch used becomes irrelevant. The dose $\tilde{D}_L(0)$ for any other desired pitch $\tilde{p}$ (including arbitrarily large clinical pitches) can subsequently be computed from $CTDI_L$ as desired using the inverse $\tilde{D}_L(0) = \tilde{p}^{-1}CTDI_L$, with no loss of accuracy.

That being said, the following tests will illustrate and experimentally confirm the robustness of the peripheral axis data thus acquired.

3.9.2 Testing the Sensitivity of the Peripheral Axis Data to Averaging Errors

Since the short ion chamber of length ℓ averages the periodic dose distribution of period b on the peripheral axes (not an issue for the central axis where the helical dose distribution is non-periodic and slowly varying), a quick test of the validity of the data is to change the pitch to see whether the reading (corrected for the pitch change) remains constant. Recall that a small acquisition pitch $p = b/nT$ is chosen to give at least a one-period average over ℓ ($b < \ell$ or $p < \ell/nT$), and using a two-period average $p \leq \frac{1}{2}(\ell/nT)$ is ultra-conservative. For the Farmer chamber ($\ell \approx 23$ mm), hence this will only be an issue for the largest $nT = 20$ mm and $nT = 40$ mm acquisitions. Table 3.9 shows the relative results obtained for the product of pitch and dose measured on the GE VCT scanner using the Farmer chamber for various pitch values. *Pitch* $\times$ *dose* should remain constant as long as the average remains robust.

Thus, our previously presented peripheral axis data in the various tables and graphs (acquired at the smallest pitch values $p \approx 0.5$) seem quite robust, with the drop in $p \times dose$

TABLE 3.9   Test of Averaging Errors on the
Peripheral Axis Using the Small Ion Chamber Method

Acquisition Pitch p	nT (mm)	$p \times D_L(0)$
0.531	20	1.000
0.969	20	0.994
1.375	20	0.950
0.516	40	1.000
0.984	40	0.935
1.375	40	0.647

Source: Dixon and Ballard, *Medical Physics* (2007).
$p \times D_L(0)$ values relative to smallest pitch acquisition.
(GE VCT scanner, $L = 250$ mm in the 400 mm body
phantom, Farmer chamber).

observed at $p \approx 1$ being even smaller than anticipated. Moreover, there is no reason even to acquire at the largest aperture $nT = 40$ mm, since the $nT = 20$ mm acquisition has already been shown to give the same $CTDI_a$ (as predicted) and a single value of $CTDI_a$ is sufficient. In fact, acquisition at $nT = 10$ mm would always guarantee at least a two-period average for any pitch $p < 1.15$ using the Farmer chamber.

3.9.3 Visualization of the Actual Measurement Field on the Peripheral Axis

These conclusions are bolstered by inspection of Figure 3.4 which shows the measured helical dose distributions for the two pitches $p = 0.938$ and $p = 0.563$ for $nT = 20$ mm, $L = 220$ mm, using the small focal spot; measured using the optically stimulated luminescence (OSL) ribbon dosimeters inserted into the peripheral axis holes in the body phantom. The large focal spot would perhaps smooth these somewhat; even so, for the smaller pitch ($p = 0.563$) the amplitude of oscillation about the mean is only about $\pm 8\%$ and is readily averaged out by the chamber length (illustrated schematically in Figure 3.4). Recall this ($p = 0.563$) technique was used for all peripheral axis acquisitions for the GE LS-16 (Table 3.4) excepting the focal spot size. In fact, at these small pitches there is little difference in the helical and axial dose distributions (Dixon et al. 2005) for the same value of b, hence an axial ("step and shoot") acquisition can also be used to acquire the data for the small ion chamber method by using scan intervals b small compared to nT.

Also included in Figure 3.4 for illustration is a $nT = 8 \times 1.25$ mm $= 10$ mm acquisition on a Lightspeed using a pitch of 0.875, which illustrates that smoothing is optimized (Dixon et al. 2005) when the gaps on the peripheral axis are filled, i.e., when $b = a(0) = a/M$, where a is the aperture and M is the relative magnification factor from the peripheral axis to the AOR (for the GE MDCT scanners, SAD $= 541$ mm, hence $M^{-1} = 0.723$ for the body phantom). From Appendix A, the appropriate aperture for this acquisition is $a = 12$ mm, hence the optimum smoothing pitch $p_s = 0.723(a/nT) = 0.87$. Although such an exact pitch selection is rarely possible, it can provide a target pitch between the one- and two-period chamber averages, although our experimental data above indicates that a one-period average is adequate.

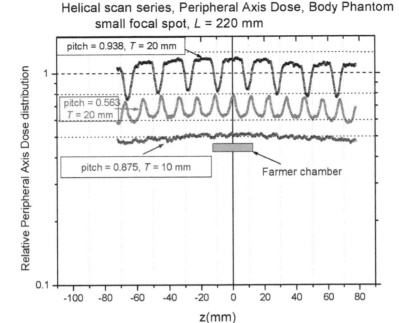

FIGURE 3.4 Peripheral axis dose distributions, body phantom, helical scans. A log scale has been chosen for the y-axis in Figure 3.4 such that equal Δy distances represent approximately the same % change. The ordinate is arbitrarily chosen for separation and ease of comparison.

(**From Dixon and Ballard,** *Medical Physics,* **2007.**)

Various other test methods such as a small shift (a few mm) in the scan-start location can also be used to verify the robustness of the peripheral axis data.

3.9.4 An Alternate Method to Circumvent the Possibility of Peripheral Axis Averaging Errors

We should point out for those still concerned about the peripheral axis acquisition, or those wanting to avoid any possibility (or even having to think about it), that a rigorous determination of $D_L(0)$ or $CTDI_L$ for the peripheral axes can be made, based on the fact that Eqs (3.1 and 3.2) represent the helical *angular average* (Dixon 2003; Dixon et al. 2005), by averaging measurements made in a ring (of at least four) symmetrically located peripheral axis points around $z = 0$. With this method, the averaging is done over θ rather than z, and thus the ion chamber length can be quite short. However, our data using the Farmer chamber would indicate that this extra effort is unnecessary.

3.10 A SUGGESTED NEW CT DOSE MEASUREMENT PROTOCOL

Determining the limiting equilibrium dose D_{eq} approached for long scan lengths, or $CTDI_\infty = pD_{eq}$, is arguably the proper goal to more realistically represent accumulated dose for clinical body scans – and preferable to $CTDI_{100}$ in that regard. Additionally, as was shown

TABLE 3.10   Measured CTDI-aperture. Values Obtained from the Product of the $nT = 20$ mm $CTDI_\infty$ Values (Tables 3.4 and 3.5) and the Corresponding Dose Efficiencies $(a/nT)^{-1}$

Scanner	Dose Efficiency $(a/nT)^{-1}$	CTDI-Aperture (mGy/100 mAs)	
		Central Axis	Peripheral Axis
GE VCT-64	0.893	8.30	12.44
GE LS-16	0.974	8.63	12.18

Source: Dixon and Ballard, *Medical Physics* (2007).

in Chapter 2, the total energy absorbed cannot be determined even for sub-equilibrium scan lengths (including 100 mm) without knowledge of $CTDI_\infty$.

We have demonstrated that $CTDI_\infty$ can be readily measured using the small ion chamber acquisition method in a phantom of sufficient length $h > L_{eq}$. One need only to use a *single* scan length $L \geq L_{eq}$ which is long enough to achieve equilibrium using only a *single* known value of the aperture a chosen small enough to provide a suitably small acquisition pitch p as discussed. In the case of the GE MDCT scanners for which the robustness of the aperture data has been established, the only dose measurement needed is the dose for the largest ($L = 380$ mm) scan length at a single aperture setting corresponding to, say, $nT = 20$ mm; from which the value of the constant $CTDI_a$ can be determined as shown in Table 3.10, and from which the entire set of $CTDI_\infty$ values for any aperture (or nT) can then be generated from Eq. (3.6) using the aperture values provided in Tables 3.8 or A1, or the dose efficiencies provided on the CT console.

Of course, the small ion chamber method does not *require* one to use the CTDI-aperture short cut, and individual measurements for every desired aperture (or nT) can be made as illustrated by our VCT data obtained at both $nT = 20$ mm and 40 mm.

(As previously noted, the $CTDI_a$ values for the GE VCT, computed using the $nT = 40$ mm data, agree with the above $nT = 20$ mm data to within 0.1% on the AOR, and within 1.0% on the periphery.)

The Table 3.10 data represents measured $CTDI_a$ values for *two different scanners* (albeit having the same model x-ray tube and equivalent bow-tie filter design), hence perfect agreement is not anticipated; nonetheless the values are in good agreement.

Given robust aperture data, a given CT scanner can be represented by a single pair of dose numbers, viz. $CTDI_a$ for the central and peripheral axes, from which the entire $CTDI_\infty$ (or even $CTDI_{100}$) tables can be generated using Eq. (3.6) for all possible collimator configurations and both focal spots as has been demonstrated for the GE MDCT scanners.

This method is quite simple, the only difficult part being the extra length and thus weight of the phantom required to obtain the desired value of $CTDI_\infty$. We had little difficulty in handling the 400 mm phantom described herein, and such a phantom if properly designed is readily manageable – particularly with modular construction. However, a water phantom might be a better choice, and recent survey data by Toth et al. (2006) suggests that a 30 cm diameter water phantom better represents an average adult body.

3.11 THE CENTRAL AXIS DOSE GAINS IN RELATIVE IMPORTANCE

The current system for approximating the planar average dose over the central scan plane (at $z = 0$) denoted by $CTDI_w$ is based on the *ad hoc* assumption that the accumulated dose in that plane $D(r)$ depends linearly on r, which results (Leitz et al. 1995) in the familiar (1/3, 2/3) weighting factors for the (central, peripheral) axes.

Since $CTDI_{100}$ is typically the only measurement made, this exclusivity may mislead one into assuming that the central/peripheral axis dose ratio in the body phantom is always about ½. However, as additional scan length is added beyond 100 mm, *the central axis dose gains on the peripheral axis* due to the larger S/P ratio at the center, such that the central/peripheral axis dose ratio increases from 0.5 at $L = 100$ mm up to 0.7 at equilibrium (see Figure 3.3).

3.12 SUMMARY OF RESULTS

A series of measurements in a 400 mm long, 32 cm diameter PMMA body phantom has illustrated the following:

1. *Good agreement between small ion chamber and pencil chamber acquisition methods.* The accumulated doses $D_L(0)$ or $CTDI_L$ values determined agreed to within ±2% at the scan lengths $L = \ell$ for both pencil chamber lengths $\ell = 100$ mm and 150 mm (Table 3.7).

2. *The commonly used phantom length of 150 mm is too short even for the measurement of $CTDI_{100}$,* producing an underestimate (compared to $CTDI_{100}$ measured in the 400 mm phantom) of 7.3% on the central axis and 1.3% on the peripheral.

3. *$CTDI_\infty$ more closely represents accumulated doses for clinically relevant body scan lengths than $CTDI_{100}$.* The approach to the limiting equilibrium dose $D_{eq} = p^{-1} CTDI_\infty$ has the form of an exponential growth curve (Figure 3.3); thus, for typical clinical scan lengths, *any overestimate of dose by $CTDI_\infty$ is much smaller than its underestimate by $CTDI_{100}$* (e.g., for $L \geq 250$ mm, the accumulated dose is ≥98% and ≥92% of D_{eq} on the peripheral and central phantom axes, respectively).

4. *The measured shortfall of $CTDI_{100}$ in predicting $CTDI_\infty$ in the body phantom* for the GE MDCT scanners at 120 kVp is given by $CTDI_\infty = 1.72 \times CTDI_{100}$ for the central axis and $1.22 \times CTDI_{100}$ for the peripheral axes, independent of nT, and in good agreement with data of other investigators.

   The relative importance of the central axis dose is increased on two fronts.

5. For clinically relevant scan lengths, the body phantom dose is more uniform than anticipated from $CTDI_{100}$ for which the center/periphery ratio is about 0.5; whereas the same ratio for the equilibrium doses ($CTDI_\infty$) is 0.7.

6. *The suggested use of aperture data to exploit the constancy of CTDI-aperture* has the potential to significantly reduce CTDI data collection time and is certainly robust (being based only upon conservation of energy).

7. *A simple method to independently verify (or determine) the appropriate active length parameter ℓ for a pencil chamber was developed* and successfully demonstrated for two different chamber models and lengths. This method is considerably simpler than the slit scanning method (Bochud et al. 2001; Jensen et al. 2006).

3.13 CONCLUSION

The efficacy and versatility of the small ion chamber acquisition method has been *decisively proven* in this extensive set of measurements. It provided accurate and robust accumulated dose $D_L(0)$ (or $CTDI_L$) values for any desired scan length; and these $CTDI_L$ values were in close agreement with those obtained using the pencil chambers of length ℓ at the appropriate $L = \ell$. There were no hidden pitfalls encountered using this new method (the transport and handling of the two-section 400 mm long phantom was not a significant problem).

Although determining the equilibrium dose (or $CTDI_\infty$) involves a longer and thence heavier body phantom; one should nonetheless strive to measure what is meaningful rather than what is expedient. It is a straightforward engineering problem to create a manageable phantom suitable for measuring the equilibrium dose. It is not practical to adapt the pencil chamber to the task of measuring $CTDI_\infty$, nor does it need to be, since we can simulate a pencil chamber of any arbitrary length using our small ion chamber methodology.

The ability to measure $CTDI_\infty$ is by no means the only reason to implement this system. Returning CT dosimetry to the realm of fundamental (and straightforward) absolute point-dosimetry using a small ion chamber and the methodology demonstrated herein is suggested as a now-proven alternative to the oft-confusing pencil chamber methodology in which one *predicts* a dose (Dixon 2006) rather than measuring it directly. A measurement of accumulated dose can be made at any point in the phantom (for any scan length); not just at the center of the scan length ($z = 0$) which is the only location at which the pencil chamber measurement and CTDI formalism allow a dose prediction.

Additionally, the pencil chamber method requires a phantom having a uniform cross-section and density along z (*shift-invariance*) – indeed the whole CTDI concept breaks down for a phantom having a longitudinal non-uniformity along z (even a simple conical shape), since this breaks the *shift-invariant* symmetry (Dixon et al. 2005; Dixon 2006); whereas the direct measurement technique using the small ion chamber can provide a valid dose measurement in any phantom *including anthropomorphic types*.

Also, the common problem of trying to match a given helical scan protocol (for which one desires the dose) with an axial scan protocol having the same aperture (in order to use the axial pencil chamber acquisition), can be eliminated by this direct measurement technique (with proper use of acquisition pitch – see Section 3.8).

Nor can the pencil chamber adapt to wide, cone-beam CT to which conventional dosimetry using the Farmer chamber is readily applied (Fahrig et al. 2006; Dixon 2006).

Whether justified or not, the fact remains that CTDI has historically been used as an indicator of patient dose even up to the present day. Thus, the value of $CTDI_\infty$ measured in an appropriate (clinically relevant) phantom, while admittedly imperfect for this task, seems to be an improvement over $CTDI_{100}$.

APPENDIX A: ILLUSTRATION OF CTDI-APERTURE CONSTANCY

Table A.1

TABLE A.1   Relative $CTDI_{100}$ -*Aperture* Values and Typical Aperture Settings for Various Collimator Configurations for the GE Lightspeed Scanner (Normalized to Unity for 16×1.25, Large Focal Spot)

Collimator Config.	Detector Width	Large Focal Spot Relative Aperture $CTDI_a$			Small Focal Spot Relative Aperture $CTDI_a$		
$n \times \Delta T$	nT(mm)	a(mm)	*Head*	*Body*	a(mm)	*Head*	*Body*
Central axis							
16×1.25	20	20.6	1.00	1.00	20.5	1.01	1.01
8×2.5	20	20.7	1.00	1.00	20.4	0.97	0.98
4×3.75	15	16.6	1.00	1.00	16.7	1.00	1.00
16×0.63	10	12.7	1.00	1.01	12.0	0.97	0.97
8×1.25	10	12.6	1.03	1.03	12.0	0.99	0.99
4×2.5	10	11.9	1.00	1.00	11.4	1.00	1.00
4×1.25	5	7.78	1.01	1.02	7.18	1.01	1.05
4×0.63	2.5	5.38	1.00	1.02	4.88	1.03	1.05
Average ± std dev =			1.01±.01	1.01±.01		1.00±.02	0.99±.03
Peripheral Axis							
16×1.25	20	20.6	1.00	1.00	20.5	1.01	1.01
8×2.5	20	20.7	1.00	1.00	20.4	0.97	0.97
4×3.75	15	16.6	0.98	1.01	16.7	1.00	1.01
16×0.63	10	12.7	0.98	0.98	12.0	0.98	0.97
8×1.25	10	12.6	1.03	0.99	12.0	1.00	0.98
4×2.5	10	11.9	1.00	0.99	11.4	0.97	1.00
4×1.25	5	7.78	0.99	1.04	7.18	1.02	1.07
4×0.63	2.5	5.38	1.01	1.02	4.88	1.04	1.07
Average ± std dev =			1.00±.02	1.00±.02		1.00±.02	1.01±.04

GLOSSARY

MDCT: multi-detector CT
ADCL: accredited dosimetry calibration laboratory
t_0: total beam-on time for an axial or helical scan series
τ: time for single 360° gantry rotation (typically $\tau = 1$ sec or less).
$N = (t_0/\tau)$: total number of gantry rotations in an axial or helical scan series
v: couch velocity for helical scans
b: generalized table advance per rotation (mm/rot)
$b = v\tau$: for helical scans; b = scan interval "I" for axial scans

$L = \upsilon t_0$: definition of total helical scan length (total reconstructed length $<L$)

$L = Nb$: generalized definition of total scan length (axial or helical)

ℓ: active length of pencil chamber

nT: total reconstructed slice width acquired in a single rotation. Also equal to the total active detector length projected at isocenter for multi-detector CT (MDCT) (often denoted by "N × T")

$p = b/nT$: generalized "pitch"

$f(z)$: single rotation (axial) dose profile acquired with the phantom held stationary

D_{eq}: limiting value of accumulated dose approached for scan lengths $L \geq L_{eq}$

L_{eq}: scan length required for dose to approach D_{eq} (denoted symbolically as $L \to \infty$)

AOR: gantry axis of rotation

a: projection of collimator aperture onto AOR (by a "point" focal spot)

N_k: ADCL global chamber calibration factor. Air kerma per unit charge (mGy/nC)

OSL: optically stimulated luminescence

PMMA: Polymethyl-methacrylate. Also known as Acrylic, Perspex, Plexiglas, etc.

LNT: linear, no-threshold theory of biological effect vs. radiation dose

REFERENCES

AAPM, 2010 Report of AAPM Task Group 111 Comprehensive methodology for the evaluation of radiation dose in x-ray computed tomography. American Association of Physicists in Medicine, College Park, MD, February 2010, http://www.aapm.org/pubs/reports/RPT_111.pdf.

AAPM, Report 204, Size Specific Dose Estimates (SSDE) in pediatric and adult body CT examinations, American Association of Physicists in Medicine (2011).

Anderson J., Chason D., Arbique G., and Lane T., New approaches to practical CT dosimetry. Abstract SU-FF-I-14: *Med Phys* 32, 1907, (2005).

Bochud F.O., Grecescu M., and Valley J., Calibration of ionization chambers in air-kerma length. *Phys Med Biol* 46, 2477–2487, (2001).

Dixon R.L., A new look at CT dose measurement: Beyond CTDI. *Med Phys* 30, 1272–1280, (2003).

Dixon R.L., Restructuring CT dosimetry—A realistic strategy for the future. Requiem for the pencil chamber. *Med Phys* 33, 3973–3976, (2006).

Dixon R.L., and Ballard A., Experimental validation of a versatile system of CT dosimetry using a conventional ion chamber. Beyond *CTDI*$_{100}$. *Med Phys* 34(8), 3399–3413, (2007).

Dixon R.L., Munley M.T., and Bayram E., An improved analytical model for CT dose simulation with a new look at the theory of CT dose. *Med Phys* 32, 3712–3728, (2005).

Fahrig R., Dixon R.L., Payne T.L., Morin R.L., Ganguly A., and Strobel N., Dose and image quality for a cone-beam C-arm CT system. *Med Phys* 33, 4541–4550, (2006).

FDA U.S. Performance standards for diagnostic x-ray systems. *Fed Regis* 49171, (August 31 1984); Govt. Printing Office.

International Electrotechnical Commission, Medical Electrical Equipment – Dosimeters with Ionization Chamber and/or Semi-Conductor Devices used in X-Ray Diagnostic Imaging, IEC, 61674 (1997).

Jensen A., Culberson W., Davis S., Micka J., and DeWerd L., Calibration of CT ionization chambers in air-kerma length. *Abstract: AAPM annual meeting*, (2006).

Leitz W., Axelson B., and Szendro G., Computed tomography dose assessment – a practical approach. *Radiat Prot Dosim* 57, 377–380, (1995).

Ma C.M., and Nahum A.E., Calculations of ion chamber displacement effect corrections for medium-energy x-ray dosimetry. *Phys Med Biol* 40, 45, (1995).

Mayajima S., The variation of mean mass energy absorption coefficient ratios in water phantoms during x-ray CT scanning. Proceedings of 10th EGS4 Users' meeting in Japan, KEK Proceedings, 74–83, (2002–18).

Morgan H.T., and Luhta R., Beyond CTDI dose measurements for modern CT scanners. *Abstract: Med Phys* 31(6), 1842, 2004, and private communication.

Mori S., Endo M., Nishizawa K., Tsunoo T., Aoyama T., Fujiwara H., and Murase K., Enlarged longitudinal dose profiles in cone-beam CT and the need for modified dosimetry. *Med Phys* 32, 1061–1069, (2005).

Nakonechny K.D., Fallone B.G., and Rathee S., Novel methods of measuring single scan dose profiles and cumulative dose in CT. *Med Phys* 32, 98–109, (2005).

Peakheart D., Rong X., Yukihara E., Klein D., McKeever S., and Ramji F., Evaluation of An OSL dosimetry system for CT quality assurance and dose optimization. *Abstract Med Phys* 33, 2211, (2003).

Seuntjens J., and Verhaegen F., Dependence of overall correction factor of a cylindrical ionization chamber on field size and depth in medium-energy x-ray beams. *Med Phys* 23, 1789, (1996).

Shope T., Gagne R., and Johnson G., A method for describing the doses delivered by transmission x-ray computed tomography. *Med Phys* 8, 488–495, (1981).

Toth T.L., Bromberg N.B., Pan T.S., Rabe J., Woloschek S.J., Li J., and Seidenschnur G.E., A dose reduction x-ray beam positioning system for high-speed multislice CT scanners. *Med Phys* 27, 2659–2668, (2000).

Toth T.L., Ge Z., and Daly M., The influence of bowtie filter selection, patient size and patient centering on CT dose and image quality. *Med Phys* 33, 2006, (2006).

CHAPTER 4

An Improved Analytical Primary Beam Model for CT Dose Simulation

4.1 INTRODUCTION

Gagne (1989) has previously described a model for predicting the sensitivity and dose profiles in the slice-width (z) direction for CT scanners; however, the wider beams of modern MDCT and cone beam scanners result in increased penumbral asymmetry and heel effect, and call into question the applicability of this earlier *flat anode* model which ignored anode-tilt, the heel effect, and the photon energy spectrum; and only applied on the axis of rotation.

The improved model described herein transcends all of the aforementioned limitations of the Gagne model, and actually produces an analytic function representing *the primary beam dose profile* (allowing extremely fast and "noise free" simulations).

Comparison of simulated and measured dose data provides experimental validation of the model, including verification of the superior match to the penumbra provided, as well as the observable effects on the cumulative dose distribution.

An original goal was the development of an analytical model for simulating the quasi-periodic helical dose distribution on the *peripheral* phantom axes (about which little information exists in the literature); this being motivated in part to facilitate the implementation of an improved method of CT dose measurement utilizing a short ion chamber originally proposed by Dixon (2003) and subsequently corroborated by several investigators (Nakonechny et al. 2005; Morgan and Luhta 2004; Mori et al. 2005). A more detailed set of guidelines for implementing such measurements is also set out.

The analytical nature of the model has also led to the formulation of some fundamental physical principles governing CT dose which have not been previously described or clearly enunciated.

4.2 PRIMARY BEAM MODEL

4.2.1 The Simple Geometric Model

It is useful to introduce the problem with a cursory review of the familiar geometric model shown in Figure 4.1. The quantity a is the collimator aperture projected onto the z-axis, magnified by the factor $M = F/F_c$. The tilted anode causes the apparent focal spot length c' to project an asymmetric penumbra c from the collimator edges onto the z-axis, with a magnification factor for the (c/c') ratio of $(M-1)$. For a "rectangular" focal spot of uniform intensity, the dose distribution on the z-axis is a flat-topped trapezoid with asymmetric sides, having a *fwhm* $\approx a$ (unless $a < c$). For a *flat anode* model (ignoring anode-tilt, with $\alpha \to \pi/2$), the penumbra and the trapezoid become symmetric.

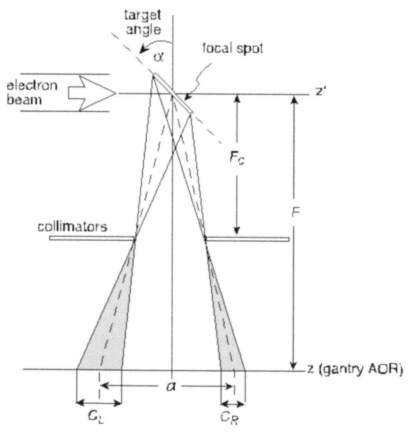

a = projected collimater aperture

c denotes penumbra width

FIGURE 4.1 Simple geometric model.

(From Dixon et al., *Medical Physics*, 2005.)

A little thought reveals another important result, namely that the primary beam profile *integral*, which is the area of the trapezoid, depends only on the product of the projected aperture a and the peak dose A_0, and is independent of the focal spot size or angle.

4.2.2 Detailed Primary Beam Model on the Axis of Rotation

The dose profile $f(z)$ resulting from a single axial rotation can be expressed, without loss of generality, as the sum of a *primary* ("uncollided") component and a *scattered* component, i.e., $f(z) = f_p(z) + f_s(z)$.

The x-ray tube in CT is oriented such that the rotational axes for anode and gantry are parallel to eliminate the torque on the anode. Figure 4.2 illustrates a planar focal spot, tilted by the anode target angle α with respect to the central ray, and having an emission intensity distribution of $\dot{S}_A(s, x')$ photons/cm^2/sec emitted with a Bremsstrahlung (BS) energy spectrum (the relative size of the focal spot has been highly exaggerated for clarity, and a dot over a quantity indicates per unit time). In order to simulate the heel effect, it is assumed that the planar source is a small distance d_0 beneath the actual anode surface which is depicted by the dashed line.

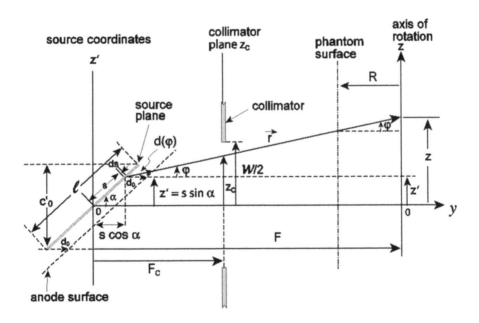

FIGURE 4.2 Geometry for the generalized primary beam model. The tilted anode (target angle α), is shown on the left, where the source (focal spot) plane is $s - x'$, with x' perpendicular to the page, and s is the focal spot length variable (with ℓ denoting the nominal focal spot length in that plane, and $c_0' = \ell \sin \alpha$ is its (optical) projection perpendicular to the central ray $y - y'$). The focal spot size is highly exaggerated for clarity. The (pre-patient) collimator plane is shown in the center, where w denotes the collimator aperture and z_c denotes a coordinate in that plane. The axis of rotation (AOR) z is shown at right, with R denoting the phantom radius.

(From Dixon et al., *Medical Physics*, 2005.)

The small width of the focal spot in the x' direction (≈ 1 mm) can be ignored in computing the dose along z, since the distance $F \gg 1$ mm, hence one need only consider a one-dimensional focal spot (line source) having an intensity per unit length along the anode surface given by,

$$\int_{-\infty}^{\infty} \dot{S}_A(s,x')dx' \equiv \dot{S}\frac{1}{\ell}g\left(\frac{s}{\ell}\right) = \left\{\begin{array}{l}\text{number photons / sec emitted} \\ \text{per unit length at } s\end{array}\right\} \tag{4.1}$$

where the chosen form of the emission intensity is a scaled function, $\ell^{-1}g(s/\ell)$, normalized to unit area over $(-\infty, \infty)$, thus $\dot{s}=$ the total number of photons/sec emitted by the focal spot. Note that this 1-D focal spot intensity function (related to the LSF) also contains the intensity sum across the focal spot width ("double ridge") dimension. The scaling parameter ℓ is representative of the *fwhm* of the intensity function $g(s/\ell)$, e.g., for a Gaussian focal spot, $g(s/\ell) = \exp[-\pi(s^2/\ell^2)]$, and $\Pi(s/\ell)$ for a "rectangular" focal spot. The emitted photons are presumed to have a Bremsstrahlung energy spectrum, such that $\dot{s}=\dot{S}(E)$, the angular dependence of which is assumed negligible over the small ($\approx 1 - 2°$) angular interval in the slice-width (z) direction.

In Figure 4.2 the vector **r** depicts a possible photon path from a source-point s to a point z on the axis of rotation (AOR), and the incremental (scalar) photon flux density at z contributed by photons emitted from ds at s having energy in dE at E is,

$$d^2\Phi = \frac{\dot{S}(E)dE}{4\pi r^2}e^{-\mu(E)d(\varphi)}\frac{1}{\ell}g\left(\frac{s}{\ell}\right)ds \tag{4.2}$$

where $\mu(E)$ is the attenuation coefficient of the anode material (W-Re alloy), and $d(\varphi)$ the path length in the anode. It is preferable to express $d(\varphi) = d_0 + \Delta d(\varphi)$ and incorporate the constant factor involving d_0 into the spectral term, such that $\dot{S}_0(E) = \dot{S}(E)e^{-\mu(E)d_0}$ now represents the thick target BS spectrum emitted in the central ray direction ($\varphi = 0$); with the remaining factor, $\exp[-\mu(E)\Delta d(\varphi)]$, representing the intensity variation relative to the central ray ("heel effect").

The probability that these emitted photons will actually reach the point z on the phantom axis is the product of the collimator transmission probability function $\Pi(z_c/w)$, where w is the aperture; and the *survival probability* of a primary photon through the beam filter and phantom material, given rigorously by $\exp[-\mu_f(E)d_f/\cos\varphi] \times \exp[-\mu_p(E)R/\cos\varphi]$, where d_f denotes the central-ray beam filter thickness and R is the phantom radius.

The incremental flux density $\Phi(z,E)dE$ is obtained by multiplying $d^2\Phi$ in Eq. (4.2) by the three transmission factors listed above, and then integrating over the entire source length s. The primary beam dose-rate profile $\dot{f}_p(z)$ is then obtained from $\Phi(z,E)dE$ by multiplying it by the fluence-to-dose conversion factor $\kappa(E) = E\left(\dfrac{\mu_{en}}{\rho}\right)$ and then performing a second integration over E, resulting in the coupled, double-integral shown below,

$$\dot{f}_p(z) = \int_E \frac{\kappa(E)\dot{S}_0(E)dE}{4\pi F^2}$$

$$\times \left\{ \int_S \frac{1}{\ell} g\left(\frac{s}{\ell}\right) e^{-\mu(E)\Delta d(\varphi)} e^{-[\mu_f(E)d_f/\cos\varphi]} e^{-[\mu_p(E)R/\cos\varphi]} \frac{\cos^2\varphi}{(1-s\cos\alpha/F)^2} \Pi\left(\frac{z_c}{w}\right) ds \right\}$$

(4.3)

in which, $r\cos\varphi = F - s\cos\alpha$, obtained from Figure 4.2, has been used to eliminate r.

In the inner integral over the source coordinate s, note that z_c and φ are both functions of s, connected by the *constraint equation* defined by the ray vector $\mathbf{r}$ in Figure 4.2, as,

$$\tan\varphi = \frac{z - s\sin\alpha}{F - s\cos\alpha} = \frac{z_c - s\sin\alpha}{F_c - s\cos\alpha}$$

(4.4)

For a cylindrical phantom, the dose rate on the axis of rotation (AOR) is independent of beam angle ($\theta = \omega t$) and time, hence the dose profile on the AOR due to a single (360°) axial rotation is obtained by multiplying Eq. (4.3) by the rotation time τ, (or simply dropping the dots).

The foregoing discussion has described the basic methodology for the model. The mathematical details of its subsequent development (including rigorous justification and accuracy estimates of approximations utilized) are developed in Appendix A; and the following simplified description is intended only as an outline of the derivation and its results. A glossary of parameters for easy reference is included.

4.2.3 Outline of Primary Beam Model Derivation

1. Small angle approximation, $\cos\varphi = 1$:

For a beam width corresponding to $nT = 40$ mm on the AOR: $\varphi_{max} = 2.4°$, $\cos\varphi_{max} = 0.9991$.

As shown in Appendix A, ignoring the $(\cos\varphi)^{-1}$ increase in attenuation path length in the phantom and bow-tie due to oblique incidence, results in an error of less than 0.1% for $nT = 20$ mm, and less than 0.3% for $nT = 40$ mm.

That is, the phantom is essentially uniformly attenuating across the slice-width and thus appears "transparent," *therefore the primary beam profile on the phantom axis of rotation will have the same relative shape as that in air with no phantom present.* The central-ray beam filter term is absorbed into the spectral term, such that the modified spectrum $\tilde{S}_0(E)$ then represents the BS spectrum on the central ray exiting the gantry port.

2. The heel effect factor $e^{-\mu(E)\Delta d(\varphi)}$ is shown in Appendix A to be essentially independent of the source coordinate s (to better than 0.2%), hence one can assume a point-source model with the result that $\Delta d(\varphi)$ depends only on z, allowing the energy integral to be expressed as the product $\rho(z) A_0$, where $\rho(z)$ represents the heel effect variation, and A_0 is the maximum possible primary dose on the phantom central axis (AOR) on the central ray ($z = 0$) as defined below,

$$A_0 = \int_E A_0(E)dE = \int_E \frac{\kappa(E)\tilde{S}_0(E)e^{-\mu_p(E)R}}{4\pi F^2}dE = \begin{Bmatrix} \text{total "emitted dose"} \\ \text{at the central axis} \end{Bmatrix} \tag{4.5}$$

For beam widths larger than the penumbra ($a \geq 2c_0 \approx 5$ mm), $f_p(0) = A_0$.

As shown in Appendix A, the heel effect function can be approximated to an accuracy of better than 0.16% as,

$$\rho(z) \cong 1 - \langle\mu\rangle d_0 \frac{z}{z_\alpha}\left(1 + \frac{z}{z_\alpha}\right) \tag{4.6}$$

where $\langle\mu\rangle$ is the weighted average of $\mu(E)$ over the spectrum $A_0(E)$, $z_\alpha = F \tan \alpha$ is the anode cutoff distance on the AOR, with the value $\langle\mu\rangle d_0 = 0.28$ empirically determined.

Thus, the complex Eq. (4.3) reduces to,

$$f_p(z) \cong \rho(z)A_0 \int_s \frac{1}{\ell}g\left(\frac{s}{\ell}\right)\Pi\left(\frac{z_c}{w}\right)ds \equiv \rho(z)\tilde{f}_p(z) \tag{4.7}$$

where $\tilde{f}_p(z)$ is defined as the primary beam axial dose profile with anode-tilt, *but without the heel effect.*

The approximations utilized in arriving at this equation were all accurate to better than 0.3%, as illustrated in Appendix A, where it is also shown that oblique phantom penetration for cone beams wider than 40 mm can be included in the model by replacing $\rho(z)$ in Eq. (4.7) by $\tilde{\rho}(z)$ from Eq. (A.14) or Glossary.

The above equation can be put into a more useful form by defining the functions,

$$c(z) = (1 - z/z_\alpha)c_0 \tag{4.8}$$

$$\bar{g}(z) = g(-z) \tag{4.9}$$

and changing the source integration variable from s to the "collimator-plane" coordinate projected on the AOR, $\xi = Mz_c$ (see Appendix A for details). The primary beam profile can then be expressed as,

$$f_p(z) = \rho(z)\tilde{f}_p(z) = \rho(z)A_0 \int_\xi \frac{1}{c(z)}\bar{g}\left(\frac{z-\xi}{c(z)}\right)\Pi\left(\frac{\xi}{a}\right)d\xi \tag{4.10}$$

This closely resembles the form of a convolution (which was the intent); however, closer inspection reveals that it cannot be expressed as such due to the variable factor $c(z)$.

It is instructive to see how Eq. (4.10) simplifies in *Flatland* where there is no anode-tilt.

Flat anode model:

If the anode-tilt is removed such that $z_\alpha \to \infty$, then from Eq. (4.8), $c(z) = c_0 = $ constant, and Eq. (4.10) can be simplified as the convolution of the reversed focal spot intensity function $g(-z/c)$ with the rectangular function $\Pi(z/a)$,

$$f_P(z) = A_0 \frac{1}{c_0} g\left(\frac{-z}{c_0}\right) \otimes \Pi\left(\frac{z}{a}\right) \tag{4.11}$$

If c_0 is small compared to a, the convolution is seen to add a symmetric penumbra of width c_0 to $\Pi(z/a)$, and $f_p(0) = A_0$ is the peak dose. The form in Eq. (4.11) can be shown to be equivalent to the corresponding equation in the Gagne model (Gagne 1989) for the exposure profile.

Eq. (4.10) can be better understood by noting that the troublesome parameter, $c(z) = (1 - z/z_\alpha)c_0$, has a simple interpretation as is shown in detail in Appendix A. Namely, its counterpart in the focal spot plane, $c'(z) = (1 - z/z_\alpha)c_0'$, represents the apparent focal spot length (optical projection) when viewed from a point z on the AOR. Reference to Figure 4.1 will reconcile its mathematical form with physical intuition.

This variation of the LSF along z is an example of *shift-variance* which breaks the symmetry required for the convolution.

An interpretation of the form of Eq. (4.10) is also useful. As $c(z)$ varies in width across the slice, the scaled focal spot intensity function (LSF) in the integral automatically maintains a constant area (integral) for any value of z. As $c(z)^{-1}$ gets larger; the function $\overline{g}(\xi/c(z))$ gets narrower to compensate, thus delivering the same dose A_0 on the axis (in the absence of the heel effect) for those values of z inside the penumbra. Inspection of Figure 4.1 is helpful in visualizing this effect. *Thus, the variation of $c(z)$ has no effect in Eq. (4.10) inside the useful beam*, and can only affect the beam in the penumbra region; producing a larger penumbra width on the negative z side than on the positive side, as previously anticipated from the geometric model. Although the expression for $f_p(z)$ in Eq. (4.10) is certainly useable, it is not readily integrable in analytical form, nor does it avail us the considerable advantages of a convolution format.

4.2.4 Convolution Approximation for the Tilted Anode

Eq. (4.10) can be accurately approximated as a convolution, while still maintaining the anode-tilt and asymmetric penumbra, which will greatly facilitate the analytical modeling of the primary beam profile. This "hat trick" is accomplished by defining separate right and left penumbra values by evaluating $c(z)$ in the center of the respective penumbra regions at $z = \pm a/2$, obtaining the "left/right" penumbra values shown below,

$$c_L = c_0\left[1 + a/2z_\alpha\right], \quad c_R = c_0\left[1 - a/2z_\alpha\right] \tag{4.12}$$

Although C_L and C_R differ by ±30% from the central ray projection c_0 for a beam width $a = 40$ mm; $c(z)$ varies slowly from its value at $a/2$ (or $-a/2$) *over each respective penumbra region* by only about ±1.5%, independent of *a*. Thus, one can approximate Eq. (4.10) for the tilted anode case as the convolution,

$$\tilde{f}_p(z) = A_0 \left[\frac{1}{c_L} g\left(\frac{-z}{c_L}\right) \otimes H(z+a/2) - \frac{1}{c_R} g\left(\frac{-z}{c_R}\right) \otimes H(z-a/2) \right] \quad (4.13)$$

where $H(z)$ is the (Heaviside) unit step function.

It is useful to check Eq. (4.13) in Flatland:
Since $\Pi(z/a) = [H(z+a/2) - H(z-a/2)]$, in the case of zero anode-tilt where $c_L = c_R = c_0$, Eq. (4.13) reduces to the flat anode model of Eq. (4.11).

4.2.5 Integral Theorem

The convolution format of Eq. (4.13) in the previous section also greatly facilitates the evaluation of its infinite integral; thereby revealing the important relationship,

$$\int_{-\infty}^{\infty} \tilde{f}_p(z)dz = \int_{-\infty}^{\infty} f_p(z)dz = A_0 a \quad (4.14)$$

in which the dose integral is shown to be proportional to the product of the total *emitted* dose A_0 and the aperture *a*. Nor does this theorem rely on our convolution approximation; but is much more fundamental. Even with the heel effect included, the expression still holds, and the physics is described in the following paragraph.

The aperture acts as the *energy gate*, controlling the amount of primary photon energy from the focal spot allowed to impinge on the phantom – this also depends on the total energy emitted by the focal spot (and thence on A_0). The integral in Eq. (4.14) actually represents the *total primary photon* energy deposited along the entire *z*-axis (in a small cylindrical volume about the axis), and remains equal to aA_0 – *independent of the size of the focal spot and penumbra* (as seen from the geometric model, where it was noted that the area of the trapezoidal beam profile remains equal to aA_0 regardless of the penumbra width). In fact, the integral remains equal to aA_0 even in the extreme limit where the aperture becomes a narrow slit $a \ll c_0$, such that the entire primary beam contains penumbra, producing a profile having a *fwhm* $\approx c_0$ (not *a*), and a peak dose $f_p(0) \ll A_0$.

A broader view is that the total amount of energy incident on the phantom is directly proportional to the aperture width *w* (and thus *a*), and also by inference to A_0; hence the energy deposited in the phantom *along any axis* by *both* primary and scatter interactions is likewise proportional to the aperture *a* and the emitted dose on the AOR A_0.

4.2.6 Model Application to the "Z-Flying Focal Spot"

A good test of the model flexibility is to apply it to the Siemens Sensation CT scanner (Siemens Medical Systems, Erlangen, DE) which utilizes magnetic steering to rapidly deflect the electron beam (Schardt et al. 2004; Kachelriess et al. 2005); toggling the focal spot position between two distinct locations along z'.

Denoting the two focal spot locations by $z' = 0 \pm \frac{1}{2}\Delta z'$, representing a total shift (Kachelriess et al. 2005) of $\Delta z' = 0.66\,mm$ (and ignoring the small "inverse square" correction produced by the concomitant shift of $\Delta y' = \Delta z' \cot \alpha$); the focal spot intensity function (LSF) can be modified for this case, using the axis projection $\Delta z = (M-1)\Delta z' = 1.22\,mm$, as,

$$LSF(z) = \frac{1}{c} g\left(\frac{z}{c}\right) \otimes \left[\frac{1}{2}\delta(z + \frac{1}{2}\Delta z) + \frac{1}{2}\delta(z - \frac{1}{2}\Delta z)\right] \qquad (4.15)$$

Replacing $c^{-1}g(z/c)$ with Eq. (4.15) in our previous equations for $f_p(z)$ [Eq. (4.13) for example], it is straightforward to show that the new axial dose profile is given by,

$$\breve{f}_p(z) = \left[\frac{1}{2} f_p(z + \frac{1}{2}\Delta z) + \frac{1}{2} f_p(z - \frac{1}{2}\Delta z)\right] \qquad (4.16)$$

which can then be modeled analytically using Eq. (4.17), shown in the next section.

The net effect is a narrowing of the central "flat" area of the dose profile (effectively broadening the penumbra region), with its infinite integral $A_0 a$ being unchanged; however, this is not quite the "end of the story," since it is necessary to increase the aperture to $a* = a + \Delta z$, in order to keep the penumbra beyond the ends of the active detector row (Kachelriess et al. 2005). This therefore results in an increase in cumulative dose (or CTDI) by a factor of $(1 + \Delta z/a)$, compared to that of the same stationary focal spot, amounting to a 6% increase in CTDI for $a \approx 20\,mm$.

4.3 EXPERIMENTAL VALIDATION OF THE PRIMARY BEAM MODEL ON THE AXIS OF ROTATION

4.3.1 Materials and Methods

Beam profiles were measured using an intensity modulated radiation therapy (IMRT) film dosimetry system in our radiation oncology center, utilizing Kodak EDR2 film (Zhu et al. 2003) (Eastman Kodak, Rochester, NY) which has a wide latitude ($OD_{max} \approx 4$) and a reduced effective-Z. The developed film is scanned using a Vidar VXR-16 Dosimetry Pro scanner (Vidar Medical Imaging, Herndon, VA) with a resolution of 0.18 mm, and controlled by RIT 113 software (Radiological Imaging Technology, Inc., Colorado Springs, CO) which also does the conversion to dose. The system is routinely calibrated to provide linearity corrections.

Primary beam profiles were obtained in air by placing a 35 × 43 cm sheet of EDR2 film on a 2.5 cm thick Styrofoam platform at isocenter (extended past the end of the couch in order to minimize backscatter); using an exposure of 120 kVp, 440 mAs with the gantry held stationary at $\theta = 0$.

4.3.2 Primary Beam Profiles: Measurement vs. Theory

Figure 4.3 shows the beam profile measured in air at isocenter and the theoretical match based on Eq. (4.17), acquired using the large focal spot on a GE Lightspeed 8-slice scanner

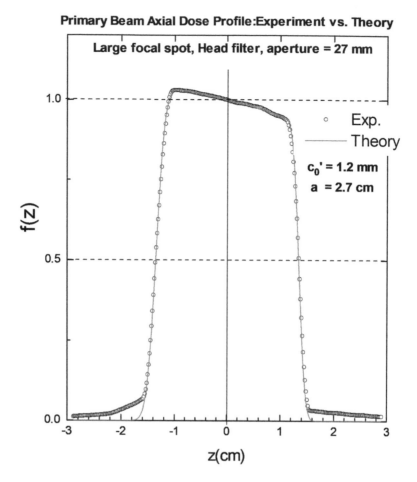

FIGURE 4.3 Measured and theoretical primary beam profiles in air. Solid line is theory and experimental data are circles.

(From Dixon et al., *Medical Physics*, 2005.)

in a service mode which allows a wide collimator aperture of $w = 8$ mm (projected aperture $a = 27$ mm) and a stationary gantry angle during exposure.

The heel effect is clearly evident. There are also small, extraneous scatter tails, probably due in part to backscatter from the Styrofoam, and possibly from scatter by the collimators and bow-tie filter. The larger "knee" on the cathode end may be due to off-focal radiation from the anode face.

In order to test our model, one ideally needs the focal spot LSF, $g(z/c_0)$, which could be obtained using a slit camera image and a microdensitometer, however that option was neither possible nor desirable. The beam profiles observed are not trapezoidal, ruling out a purely rectangular focal spot. In fact, the focal spot LSF, which includes the integrated intensity across the width dimension, is unlikely to be purely "rectangular" or purely Gaussian in shape; however, a Gaussian model (also used by Gagne) will be assumed since it provides a convenient analytical function for $f_p(z)$. Substituting the Gaussian $g(z/c) = \exp[-\pi(z^2/c^2)]$ into the convolution approximation of Eq. (4.13), we obtain,

$$f_p(z) = \rho(z)A_0 \left\{ \frac{1}{2} erf \left[\frac{\sqrt{\pi}}{c_L} \left(\frac{a}{2} + z \right) \right] + \frac{1}{2} erf \left[\frac{\sqrt{\pi}}{c_R} \left(\frac{a}{2} - z \right) \right] \right\} = \rho(z)\tilde{f}_p(z) \quad (4.17)$$

where $erf(x)$ is the error (or probability) function.

The GE Lightspeed scanners have dual focal spots with nominal *length* specifications of $(0.6, 0.9 \text{ mm})$, [IEC 336/93 standard]. As a practical estimate for c_0', a *Siemens Star* measurement was made with resulting focal spot *lengths* of $(0.63, 1.1 \text{ mm})$ for the small and large focal spots (casting doubt on their Gaussian "purity"). Values of $c_0' = (0.65, 1.2 \text{ mm})$ were found to work well in our Gaussian model, and are used for all the simulated beam profiles to follow.

A very good *match* to the primary beam data as shown in Figure 4.3, using Eq. (4.17), was obtained using $c_0' = 1.2 \text{ mm}$, and the actual value of the aperture set, $a = 2.7 \text{ cm}$.

The heel effect parameter used, $\langle \mu \rangle d_0 = 0.28 \ (\pm 0.02)$, was established empirically by "deconvolving" the profile data (by subtracting the penumbra), and fitting the slope using Eq. (4.6) for $\rho(z)$.

Figure 4.4 shows the primary beam profiles similarly obtained in air on a GE LS 16-slice scanner using the small focal spot for both a $16 \times 1.25 \text{ mm}$ ($nT = 20 \text{ mm}$) and a $4 \times 2.5 \text{mm}$ ($nT = 10 \text{ mm}$) collimator configuration. Good *matches* were obtained as illustrated in Figure 4.4 using $c_0' = 0.65 \text{ mm}$, and apertures of $a = 20.6 \text{ mm}$, and $a = 11.4 \text{ mm}$, consistent with expected values in Table 4.1.

Note that this matching of theory and experiment is not just a two-parameter fitting game, since the choice of aperture is strictly constrained by Table 4.1; and, for a dual focus CT tube, only two distinct values of c_0' may be chosen. In addition, for the larger beam widths, $a > 2c_0$, the *fwhm* depends only on a, independent of c_0'.

4.3.3 Tilted vs. Flat Anode Challenge

Is the tilted anode model worth the trouble; or can just as good a match be achieved with the *flat anode* model of Eq. (4.11)? A fit to the $nT = 20 \text{ mm}$ profile in Figure 4.4 was attempted assuming no anode-tilt (symmetric penumbra). The *flat anode* match in both right and left penumbra regions was clearly inferior to that obtained with our tilted anode model using the same values of a and c_0' in both cases. Moreover, *the flat anode fit could not be further improved by any adjustment of the values of c_0' or the aperture a, and clearly does not predict the readily observable heel effect.*

4.4 SENSITIVITY PROFILE

Although the main emphasis of this paper is the dose profile, it is relatively straightforward to adapt our model from dose to image acquisition (detection).

The heel effect is compensated for by detector balancing during calibration, and hence is not a factor in the image acquisition.

If the penumbra falls outside the active detector row, as in MDCT, the effect of the focal spot on the sensitivity profile (Gagne 1989; Hsieh 2004) is much smaller than on the dose

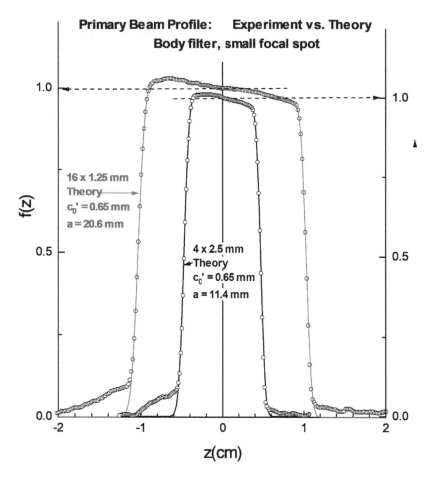

FIGURE 4.4   Primary beam profiles on the axis of rotation. GE 16-slice scanner using small focal spot. Solid lines are theoretical predictions and circles represent experimental values. Vertical scales are offset for separation and clarity.

(From Dixon et al., *Medical Physics*, 2005.)

profile; therefore the asymmetry of the sensitivity profile produced by anode-tilt will be much harder to observe, and the Gagne model (Gagne 1989) is probably an adequate approximation.

4.5 SCATTER MODEL

Scatter makes a major contribution to the cumulative dose (or CTDI) and, in fact, dominates the primary by a large factor on the central axis of the phantom.

The scatter contribution to the single-slice dose $f(z)$ can be described by the convolution of a scatter impulse response function (LSF) with the primary beam function, based solely on the symmetry argument *that the scatter along a longitudinal axis in a uniform, cylindrical phantom is a linear, shift-invariant phenomenon (Barrett and Swindell 1981).* That is, if a narrow slit (or knife-edge) beam of primary photons (an impulse) impinges on the phantom perpendicular to the z-axis at any location z', the local scatter impulse-response function about z', $LSF(z - z')$, is independent of the location of z' along the axis; and is therefore

TABLE 4.1   Typical Collimator Apertures for the GE Lightspeed
Family of CT Scanners, as Projected onto the Axis of Rotation

		Large Focal Spot	Small Focal Spot
$N \times \Delta T$ (mm)	nT(mm)	Aperture a(mm)	Aperture a(mm)
16×1.25	20	$20.6 \pm 0.5\%$	$20.5 \pm 0.7\%$
8×2.5	20	20.7	20.4
4×5	20	21.1	20.2
$4 \times 3,75$	15	$16.6 \pm 0.7\%$	$16.7 \pm 0.7\%$
16×0.63	10	12.7	12.0
8×1.25	10	12.6	12.0
4×2.5	10	$11.9 \pm 1.0\%$	$11.4 \pm 1.3\%$
8×0.63	5	7.90	7.13
4×1.25	5	$7.78 \pm 2.6\%$	$7.18 \pm 2.7\%$
4×0.63	2.5	5.38	4.88

Source: Dixon et al., *Medical Physics* (2005).

These values were obtained from one particular scanner in the factory
during calibration (Toth), and the expected variability (std.
dev.) from scanner-to-scanner is listed (Toth et al. 2000).

shift invariant. Since any primary beam function can be written as a linear superposition of impulses of varying strength, this leads directly to the convolution representation for the scatter contribution to the axial profile $f_s(z) = LSF(z) \otimes f_p(z)$. Expressing the scatter LSF in the same scaled and normalized format used previously for the focal spot intensity, where the scatter-to-primary ratio η is the scatter response to a unit-strength, primary beam impulse, and λ is the *fwhm* of the LSF, then,

$$f_s(z) = LSF(z) \otimes A_0\Pi(z/a) = A_0\eta\left[\frac{1}{\lambda}\hat{h}\left(\frac{z}{\lambda}\right)\right] \otimes \Pi(z/a) \qquad (4.18)$$

(The primary penumbra has been ignored since its width $c \ll \lambda$).

The importance of being able to express the scatter contribution $f_s(z)$ as such a convolution is that this form ensures that its infinite integral ($\eta\, aA_0$) is also proportional to the product aA_0 (aperture a and emitted dose A_0), *independent of any details regarding the functional form of the LSF(z).* Thus, the total scatter plus primary energy deposited along the axis, i.e., the integral of the total axial dose profile $f(z) = f_p(z) + f_s(z)$

$$\int_{-\infty}^{\infty} f(z)dz = A_0a(1+\eta) \qquad (4.19)$$

is proportional to A_0a. It is also clear from Eq. (4.19) that η represents the scatter-to-primary ratio of the respective contributions to the equilibrium dose (or CTDI). Recall that $A_0 = f_p(0)$ for typical CT beam widths wider than the penumbra.

Having described the general properties of the scatter contribution, from which much of the interesting physics derives, the exact form of the LSF is not essential to the objective of investigating the quasi-periodic dose distributions, since it is the primary beam function which is responsible for the oscillations of the cumulative dose. Scatter, in fact, improves the smoothness of this dose by effectively reducing the amplitude of the oscillations [by a factor of $(1+\eta)^{-1}$ at equilibrium].

For the body phantom, $\eta = 13$ on the central axis and $\eta = 1.5$ for the peripheral (Dixon and Boone 2011).

4.5.1 Defining CTDI-Aperture

It is useful to re-introduce the fundamental quantity, dubbed *CTDI-aperture*, to illustrate some basic concepts. If the line integral in Eq. (4.19) is divided by the aperture a (rather than nT as for CTDI), its value remains constant, independent of a for all detector configurations nT ("$N \times \Delta T$"), namely,

$$CTDI_a = \frac{1}{a} \int_{-\infty}^{\infty} f(z)dz = A_0(1+\eta) = \text{constant} \qquad (4.20)$$

It is also constant for "free-in-air" ($CTDI_{air}$) measurements of the primary beam.

That is, $CTDI_a$ describes a basic constancy deriving from the energy deposited, and such quantities are always of interest in physics. It also represents a lower limit (or baseline) for the conventional CTDI values, since $CTDI = (a/nT)CTDI_a$, and $a/nT > 1$ for MDCT (in fact a/nT is a direct measure of "over-beaming")

Its constancy has been verified to within a few percent (using the $CTDI_{100}$ values for a GE 16-slice scanner) for both axes in both phantoms for both focal spots, over the complete range of a in Table 4.1 as illustrated in Chapter 3 (the conventional CTDI varies by more than a factor of two over this same range of a). $CTDI_a$ can be measured in the same manner as the CTDI, dividing the measured integral by a rather than nT. In fact, using a measurement of $CTDI_a$ at only one value of a, and using a/nT from Table 4.1, one can generate the complete set of CTDI values for detector configurations $N \times \Delta T = nT$.

The constancy of $CTDI_a$ has really nothing to do with the *fwhm* of the beam. In the first place, a is *not* equal to the *fwhm* of the total beam profile, nor does it always equal the *fwhm* of the *primary* beam profile – indeed the *fwhm* has little physical significance or relevance. Rather the constancy of $CTDI_a$ is a direct result of energy conservation.

4.6 EXTENSION OF THE PRIMARY BEAM MODEL TO THE PERIPHERAL PHANTOM AXES

Theoretical treatments in CT rarely seem to venture into the largely unexplored region of the peripheral phantom axes (or peripheral detectors), which are displaced from both the axis of rotation (AOR) and laterally from the central ray. However, predicting the cumulative dose distributions for these axes, where the nature of the helical dose distribution is fundamentally different from that on the central axis, is of prime interest – both for completeness and for achieving a better understanding of the measurement problem.

Figure 4.5 illustrates the geometry for the problem, where P denotes a peripheral z-axis normal to the page, located at a radius R_0 from the axis of rotation. For convenience, an equivalent picture is used in which the x-ray tube is shown as stationary with the phantom being rotated.

It is clear from Figure 4.5 that the dose rate on a peripheral axis is a function of time (or angle θ), due primarily to the variable attenuation path length in both the "bow-tie" filter and the phantom, such that the maximum dose rate occurs at $\theta = 0$.

Our previous model and equations for the dose rate $\dot{f}(z,t)$ are valid for any fixed value of the beam angle $\theta = \omega t$ if attenuation path lengths and distances are adjusted correctly.

It is convenient to express the dose rate as a function of θ rather than time, viz., $\dot{f}(z,t) = \tau^{-1} f(z,\theta)$, such that the "single-slice" axial dose profile generated in a single rotation of 2π without phantom motion is given by,

$$f(z) = \frac{1}{2\pi} \int_{-\pi}^{\pi} f(z,\theta)d\theta \qquad (4.21)$$

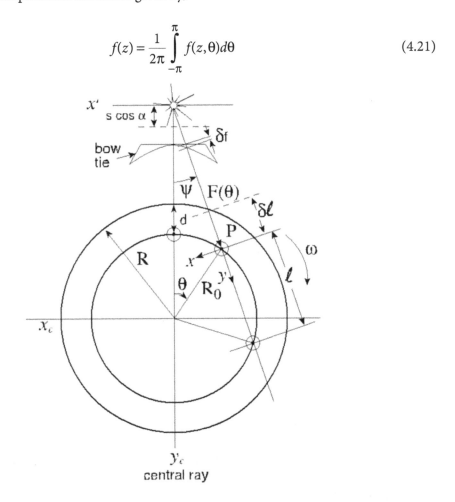

FIGURE 4.5 Geometry for model extension to peripheral axis P. The peripheral axis P (perpendicular to the page) is shown in three representative locations relative to the x-ray source during a single rotation. Path lengths in attenuating materials ℓ, $\delta\ell$ and δf are also indicated.

(From Dixon et al., *Medical Physics*, 2005.)

Likewise, the emitted dose rate transforms as $\dot{A}(t) = \tau^{-1} A(\theta)$, with the total emitted dose accumulated on the axis P in one rotation equal to an integral of $A(\theta)$ of the above form and denoted by A (see Table B.1, Appendix B).

4.6.1 Primary Beam Model for the Peripheral Axes

The mathematical details of the derivation are given in Appendix B, however, the physics of the problem can be understood as follows.

The peripheral axis P at a given instant of time has rotated through an angle θ, and is offset laterally from the x-ray tube central ray by an angle $\psi = \psi(\theta)$, which is the angle adopted as defining the "pseudo central-ray" (y-axis) for the peripheral axis coordinate system (x, y, z) shown in Figure 4.5.

A complication is that the anode and focal spot plane (s, x') are tilted toward the true central ray of the x-ray tube at $\psi = 0$, thus the focal spot plane (from the oblique point of view of P) is rotated about z' by an angle ψ, thereby making an angle of $(90° - \psi)$ with the y axis of P. This rotation does not affect the z-components of the focal spot; however, if one imagines a rectangular focal spot area, the focal spot when viewed from P will be distorted, appearing as a parallelogram having the same vertical width along z, but foreshortened along x by $\cos\psi$; and also along y by $\cos\psi$, the latter causing the anode target angle α to appear slightly larger from P. Use of a central line source is readily justified as before, and the focal spot intensity function to be utilized $[g(z/c)]$ will be the same as that used for the AOR, since the z-components and total source strength are unchanged.

The net effect of this obliquity on the formulae previously derived for the axis of rotation will be the appearance of the factor $\cos\psi$ attached to anything having a y-component, while leaving the z-components unaffected as illustrated in Appendix B.

In addition, the focal-to-axis distance F for axis P is $F(\theta)$, hence numerous quantities which depend on magnification or projection will also become functions of θ (or ψ), such as the projected aperture $a \to a(\theta)$.

The collimator plane which is perpendicular to the true central ray, likewise appears rotated by $(90° - \psi)$, such that the source-to-collimator distance is increased slightly along y to $F_c(\theta) = F_c/\cos\psi$.

These effects, however complicated they may seem, are in fact quite small, since the maximum values of ψ geometrically possible are small (7.4° for the head phantom and 16° for the body phantom, such that $\cos\psi = 0.99$ and 0.96, respectively); moreover, as will be seen, increased attenuation in the phantom and bow-tie filter (primarily the latter), reduces the relative contribution to the total $f(z)$ at the larger values of ψ, effectively limiting ψ to about 10° for the body phantom.

Following step-by-step the previous derivation for the axis of rotation illustrated in Appendix A, it is easy to show that the results for the peripheral axis will be relatively minor modifications to the equations derived for the AOR, as shown in Table B.1 in Appendix B.

For the convolution approximation, where a and z_α denote the values on the axis of rotation, we have,

$$c_L(\theta) = c_0(\theta)\left[1 + a\cos\psi / 2z_\alpha\right], \quad c_R(\theta) = c_0(\theta)\left[1 - a\cos\psi / 2z_\alpha\right] \tag{4.22}$$

$$\tilde{f}_p(z,\theta) = A(\theta)\left[\frac{1}{c_L(\theta)}g\left(\frac{-z}{c_L(\theta)}\right)\otimes H(z+a/2) - \frac{1}{c_R(\theta)}g\left(\frac{-z}{c_R(\theta)}\right)\otimes H(z-a/2)\right] \quad (4.23)$$

Assuming a Gaussian focal spot, and including the heel effect, this becomes,

$$f_p(z,\theta) = \rho(z,\theta)A(\theta)\left\{\frac{1}{2}erf\left[\frac{\sqrt{\pi}}{c_L(\theta)}\left(\frac{a(\theta)}{2}+z\right)\right] + \frac{1}{2}erf\left[\frac{\sqrt{\pi}}{c_R(\theta)}\left(\frac{a(\theta)}{2}-z\right)\right]\right\} \quad (4.24)$$

Figure 4.6 shows "in air" profiles obtained from film scans at $\psi = 0°$, $\psi = 7°$, and $\psi = 9°$ (ψ is shown in Figure 4.5). The theoretical *matches* shown in Figure 4.6 *using Eq. (4.24)* with the parameters in Table B.1 appear satisfactory, with no unpleasant surprises.

While it is still possible to apply the convolution to the primary beam *dose rate* function $f_p(z,\theta)$ at a given angle, the total primary dose profile, obtained by summing the contributions over θ, *cannot be expressed as a convolution* due to the angular dependence of $c_L(\theta)$

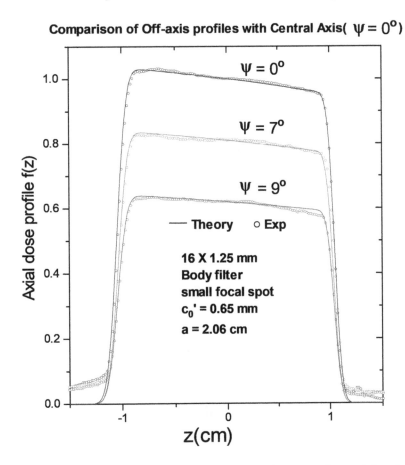

FIGURE 4.6 Test of model assumptions for the primary beam on the peripheral axes. The solid lines are the theoretical predictions using Eq. (4.24) with the parameters shown in Appendix B, Table B.1. The circles represent experimental film data.

(From Dixon et al., *Medical Physics*, 2005.)

and $c_R(\theta)$ as well as $a(\theta)$ – not even for the *flat anode* model. Nonetheless, it is straightforward to integrate Eqs (4.23 or 4.24), given the relative variation of the emitted dose rate $A(\theta)$ as a function of beam angle θ.

4.6.2 Determination of the Angular Dependence of the Primary Beam Dose Rate $A(\theta)$

Reference to Figure 4.5 illustrates the attenuation path lengths to the peripheral axis P through the bow-tie filter material $\delta f(\theta)$, and through the phantom $\ell(\theta)$ and $\delta\ell(\theta)$; with only $\delta\ell(\theta)$ contributing for $\theta \leq \theta_c = \cos^{-1}(R_0/F)$, at which angle the ray to P in Figure 4.5 is tangent to the circle of radius R_0. The geometric relationships between ψ, θ, $\delta\ell$, and ℓ required to calculate attenuation path lengths in the phantom are given in Table B.1 for completeness. While it is straightforward to evaluate $A(\theta)$ using the integral expression in Table B.1, and the incident spectrum, $\tilde{S}_0(E)$, given the bow-tie filter composition and thickness $\delta f(\psi)$ relative to that on the central ray; it is a much simpler matter to directly measure the bow-tie transmission profile. It should also be noted that the function $A(\theta)$ to be determined is the *primary beam* intensity function, which should ideally be measured under scatter-free conditions; however, this is not possible for beam angles larger than $\theta_c = \cos^{-1}(R_0/F)$. Fortunately, due to the rapid falloff of $A(\theta)$ with θ in Figure 4.7, most of the dose on a peripheral axis (97% for body and 83% for head) is contributed by beam angles $|\theta| \leq \theta_c \approx 75°$.

Figure 4.7 illustrates relative dose rate functions $A(\theta)$ for the head and body phantoms, measured using several different methods.

A small volume (0.3 cm³) ion chamber (PTW model 30-316, PTW Freiberg, DE) taped inside a hollow, 1 cm thick annulus was used to measure $A(\theta)\exp[-\mu_p\delta\ell(\theta)]$ in order to include the attenuation by $\delta\ell(\theta)$ in the measurement, without adding the significant scatter that a complete phantom would contribute.

Surprisingly, the results were very close to a set of dose integral measurements made in the head and body phantoms using a 10 cm long pencil chamber at fixed gantry angles, these also being included in Figure 4.7.

4.7 MODEL PREDICTIONS FOR CUMULATIVE DOSE DISTRIBUTIONS

4.7.1 Transforming the Helical Dose Equation on a Peripheral Axis into an Axial Format

As was previously shown in Chapter 2, the accumulated helical dose $D(z)$ at a point z on a longitudinal axis is obtained by integrating the instantaneous (traveling) dose rate profile $\dot{f}(z - \upsilon t, t)$ over the total beam-on time $(-\frac{1}{2}t_0, +\frac{1}{2}t_0)$ as it moves past a fixed point z in the phantom at velocity υ; the integral of which, with the following variable changes $\theta = \omega t$, and $\tilde{z} = \upsilon t$ (connected by $\tilde{z} = b\theta/2\pi$), becomes,

$$D(z) = \frac{1}{b} \int_{-L/2}^{L/2} f(z - \tilde{z}, \theta)d\tilde{z} \tag{4.25}$$

where $L = \upsilon t_0$ is the scan length, and $b = \upsilon\tau$ is the table advance per rotation.

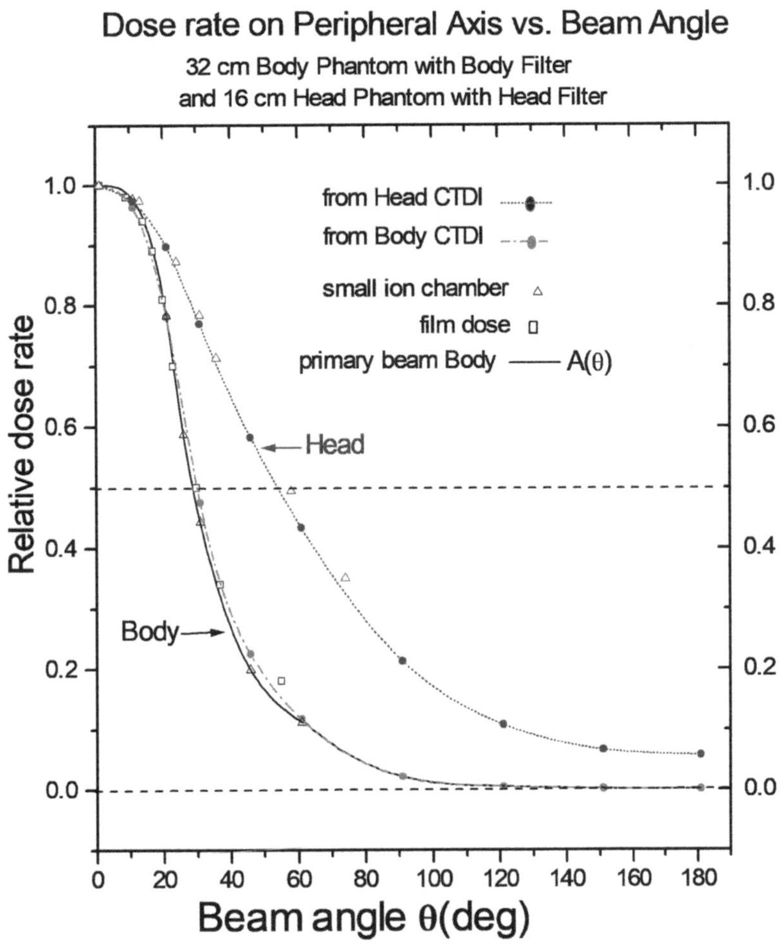

FIGURE 4.7   Relative primary dose rate on a peripheral axis vs. beam angle.

(From Dixon et al., *Medical Physics*, 2005.)

This integral can be broken up into a summation of incremental contributions for each *helical* rotation, and forced into the same form as the cumulative dose equation for axial scans, viz.,

$$D(z) = \sum_{n=-J}^{J} \bar{f}(z - nb) = \bar{f}(z) \otimes \sum_{-J}^{J} \delta(z - nb) \qquad (4.26)$$

where the total number of rotations $N = (2J+1)$ corresponds to the helical scan length $L = Nb$ with $b = \upsilon\tau$. This requires the function $\bar{f}(z)$ to have the form,

$$\bar{f}(z) = \frac{1}{2\pi} \int_{-\pi}^{\pi} f\left(z - \frac{b\theta}{2\pi}, \theta\right) d\theta \qquad (4.27)$$

By comparison of Eq. (4.27) with Eq. (4.21) for axial scans, it can be seen that $\bar{f}(z)$ represents the average dose profile along z resulting from a single *helical* rotation, during which θ varies over $(-\pi, \pi)$ and $\tilde{z} = b\theta/2\pi$ varies over $(-b/2, b/2)$, making the summation in Eq. (4.26) appear almost obvious in retrospect.

If $f(z)$ is replaced by $\bar{f}(z)$ and b with $b = \upsilon\tau$ in any equation or expression previously applicable to axial scans, the equation will likewise apply to and produce the same quantity for a *helical scan series* on the peripheral axes.

The width of $\bar{f}(z)$ increases as $b = \upsilon\tau$ (or pitch b/nT), since its components $f(z,\theta)$ are shifted laterally by $b\theta/2\pi$, hence it will be broader than $f(z)$ by an amount depending on the pitch. The difference between $\bar{f}(z)$ and $f(z)$ is small for small generalized pitches $b/nT \leq 1$ in the body phantom. Their difference is smaller in the body phantom compared to the head, since $A(\theta)$ exhibits a sharper cutoff with θ therein (Figure 4.7).

4.7.2 Basic Equations Describing the Accumulated Dose

The common formalism derived in the previous section allows the cumulative dose for both helical and axial scan series to be described by only two basic equations as shown next, where the scan length is $L = Nb$, with $N = (2J+1) =$ total number of rotations, and $b =$ table advance per rotation ($b = \upsilon\tau$ for helical, or $b =$ scan interval for axial).

1. Quasi-periodic dose distribution of period b

$$D(z) = \sum_{n=-J}^{J} f(z-nb) = f(z) \otimes \sum_{-J}^{J} \delta(z-nb) \qquad (4.28)$$

Applicable to: *Axial* scans – any phantom axis
   Helical scans – *peripheral* axes if $f(z)$ is replaced by $\bar{f}(z)$.
2. Smooth (non-periodic) dose function

$$D_S(z) = \frac{1}{b} f(z) \otimes \Pi(z/L) \qquad (4.29)$$

and its value at $z = 0$

$$D_S(0) = \frac{1}{b} \int_{-L/2}^{L/2} f(z)dz \qquad (4.30)$$

Applicable to:

1. *Actual helical* dose-*central* axis (the helical dose is naturally smooth on the AOR)

2. *Angular average* of the quasi-periodic *helical* dose on the *peripheral* axes [average at a fixed value of z over all peripheral axes $(-\pi \leq \theta_{axis} \leq \pi)$]

3. *Longitudinal average* of the quasi-periodic dose, i.e., "running mean" over $(z - b/2, z + b/2)$ for:

   Axial scans – *any axis*

Helical scans – *peripheral* axes if $f(z)$ is replaced by $\bar{f}(z)$.

It is the quantity in Eq. (4.30) which is typically "measured" in CT (called the MSAD when representing a longitudinal average about $z=0$). When the scan length L becomes large enough to encompass the entire scatter tails of $f(z)$, the cumulative dose approaches an equilibrium value D_{eq} indicated symbolically in Eq. (4.30) by setting $L=\infty$, thus $CTDI_\infty = (b/nT)\,D_{eq}$.

It is easy to show that the *infinite* integrals of $\bar{f}(z)$ and $f(z)$ are equal, and also logical since they are both built up from the same basic components $f(z,\theta)$; thus the longitudinal and angular averages listed above all converge to a single value where the local dose has reached equilibrium, e.g., the helical and axial MSAD become equal at equilibrium.

These two equations [Eqs (4.28 and 4.29)] also result in the universally applicable equation representing the total energy deposited,

$$\int_{-\infty}^{\infty} D(z)dz = N \int_{-\infty}^{\infty} f(z)dz = LD_{eq} \qquad (4.31)$$

which is proportional to the *infinite* integral of $f(z)$, and thus may be underestimated by the DLP (based on $CTDI_{100}$) for the wide beams of MDCT (Dixon 2003).

4.7.3 Smoothing Conditions for the Quasi-Periodic Cumulative Dose

It is possible to show that for axial scans on the AOR, if the *flat anode* model is utilized (no anode-tilt or heel effect); when the scan interval $b \to a$, *the normally oscillatory axial dose distribution will collapse to a perfectly smooth dose distribution of amplitude $CTDI_a$.* This results from the convolution format of Eq. (4.11), as can be shown by substituting Eq. (4.11) with $a=b$ into Eq. (4.28). "Perfect smoothing" also occurs at $b=a, a/2, a/3, a/4...$. This case presents a good test of the relevance of our model, i.e., does the inclusion of anode-tilt and the heel effect have any significant effect on the cumulative dose distribution? This will be readily evident in this "perfect smoothing" example. Figure 4.8 shows the *primary-dose* distribution predicted by our model resulting from five axial slices having $a=20$ mm, $c_0'=1$ mm, and spaced with the ideal smoothing interval $b=a$.

It is seen that both anode-tilt and the heel effect have a measurable impact on the cumulative axial primary beam dose distribution, and "perfect smoothing" on the AOR is no longer possible for axial scans, regardless of the value of b chosen. Nonetheless, $b=a$ still results in the optimal smoothing condition in this case, and *inclusion of scatter will of course, dampen these excursions.* The *helical dose on the AOR* is, however, always naturally smooth for any pitch.

On the peripheral axes, even with the *flat anode* model, "perfect smoothing" will exist only for one angular component $f_p(z,\theta)$ of the total primary beam – the one having $a(\theta)=b$. However, since $F(\theta)$ and hence $a(\theta)$ vary slowly over the small angular range about $\theta=0$ where $A(\theta)$ is appreciable (see Figures 4.5 and 4.7), the *dose-weighted* average of $a(\theta)$ with respect to $A(\theta)$ is found to be very close to its minimum value $a(0)=a(F-R_0)/F$; namely, $\langle a \rangle = 1.06\,a(0)$ for *both* the body and head phantoms.

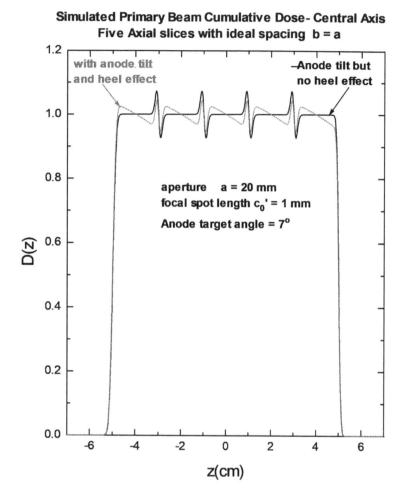

FIGURE 4.8 Departure from the "perfect smoothing" prediction of the *flat anode* model for axial scans on the axis of rotation due to inclusion of anode-tilt and the heel effect.

(From Dixon et al., *Medical Physics*, 2005.)

Thus, a reasonable first estimate for a value of *b* to achieve optimized smoothing on the peripheral axes would seem to be a value $b \approx \langle a \rangle$ and model simulations suggest *that the optimal smoothing value is b = 1.05 a(0) for axial scans and b = a(0) for helical scans.*

4.7.4 Simulation of the Accumulated Dose on the Peripheral Axes

The single-rotation, primary beam dose profiles $\bar{f}_p(z)$ or $f_p(z)$ are generated by substituting Eq. (4.24) into Eq. (4.27) [or Eq. (4.21) for axial], and integrating over θ. The primary contributions are then summed over *N* total rotations in Eq. (4.28) to obtain the accumulated primary dose component for a complete scan series. It is actually only the broad scatter-component which "accumulates" by continually adding to the central plane dose $D_L(0)$ as *N* increases, until the equilibrium dose is attained.

For expedience, the scatter background was simulated using a basic Gaussian LSF $= \eta \exp(-\pi z^2 / \lambda^2)$ which, when convolved with $A_0 \Pi(z/a)$ in Eq. (4.18), gives,

$$f_s(z) = \eta A_0 \left\{ \frac{1}{2} erf\left[\frac{\sqrt{\pi}}{\lambda} \left(\frac{a}{2} + z \right) \right] + \frac{1}{2} erf\left[\frac{\sqrt{\pi}}{\lambda} \left(\frac{a}{2} - z \right) \right] \right\} \tag{4.32}$$

which is substituted into Eq. (4.28); summed over N rotations; and added to the primary component in order to obtain the total accumulated dose.

Actually, the sum of two Gaussians having ($\eta = 1$, $\lambda = 40$ mm) and ($\eta = 0.4$, $\lambda = 200$ mm) were required to approximate the scatter background of the experimental distributions observed. In Chapter 5, analytical equations for the scatter component are derived *based on a Monte Carlo simulation* of the LSF response to a knife-edge primary beam (Dixon and Boone 2011).

4.7.5 Experimental vs. Simulated Accumulated Dose Distributions

The comparisons are limited to only a few cases as proof of concept due to space considerations. In order to experimentally map the accumulated peripheral axis dose $D(z)$ for a scan series, a sheet of EDR2 film was wrapped around the surface of the standard 32 cm body phantom, and sandwiched between the surface and a 1 cm thick Perspex annulus slid over the phantom. This, of course, simulates the dose on the peripheral axis of a 34 cm diameter phantom, however, the simulation model was changed to accommodate the larger phantom by increasing R and R_0 in the Table B.1 equations and adjusting $A(\theta)$.

Dose measurements for various scan series were made on the GE 8-slice scanner.

Figure 4.9 shows a comparison for a 8×1.25 mm ($nT = 10$ mm) helical scan series consisting of $N = 21$ rotations on the (34 cm) body phantom, large focal spot, and a pitch $p = (b/nT) = 0.625$ about halfway between the first and second order optimal smoothing pitches – *chosen to produce some significant structure to challenge the model* (with significant primary beam overlap).

The overlapping primary profiles produce the dose "spikes" observed in Figure 4.9. The model simulation is seen to provide a reasonable prediction of this "spiked" cumulative dose.

Figure 4.10 shows a helical scan series performed using the same configuration (8×1.25 mm) with a larger pitch $= 0.875$, using the small focal spot instead ($c_0' = 0.65$ mm) and $a = 1.20$ mm from Table 4.1. Since $a(0) = 0.845$ mm; the pitch of 0.875 used is only slightly above that required for optimum smoothing, resulting in the small "valleys" due to the primary beam gaps (accentuated by the small focal spot use).

The simulation prediction is reasonably good; however, it exhibits some small dips in the profiles not observed in the experimental data, which are actually due to small positive excursions at the slice edges produced by overlap of the tails on the single-rotation profiles. Using a function $A(\theta)$ resulting from an idealized bow-tie filter assumption, which has a sharper "cutoff" with θ than shown in Figure 4.7, the cupping disappears and the

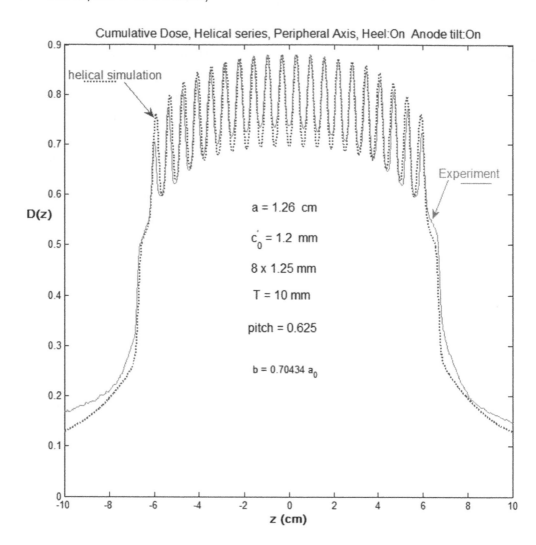

FIGURE 4.9 Comparison of experimental and simulated cumulative dose distributions for a helical scan series on a big (34 cm) body phantom using a very fine pitch = 0.625. The dose will smooth if the pitch is increased to the optimal value of $b/T = a(0)/T = 0.89$; or reduced to half that (pitch = 0.45). The solid line represents the experimental film data.

(From Dixon et al., *Medical Physics*, 2005.)

simulation fit becomes "near perfect"; however, another cause might be the Gaussian focal spot assumption.

4.8 CUMULATIVE DOSE (OR CTDI) MEASUREMENTS USING A SMALL ION CHAMBER

A direct measurement of the average cumulative dose $D(0)$ [Eq. (4.28)] can be made by placing a short (<3 cm) ion chamber at $z = 0$, and executing a helical or axial scan series of length $L = Nb$. This is equivalent to a dose integral measurement made using a pencil

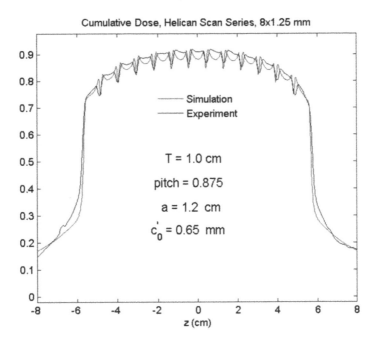

FIGURE 4.10   Comparison of experimental and simulated cumulative dose distributions for a helical scan series on a big (34 cm) body phantom for the small focal spot using a pitch = 0.875 (just above the predicted optimum smoothing pitch of 0.845). The simulation curve exhibits small central dips ("cupping") not seen in the experimental data.

(From Dixon et al., *Medical Physics*, 2005.)

chamber of length L. The robustness of this method (as well as its advantage) has been demonstrated in detail in Chapter 3.

4.9 SUMMARY AND CONCLUSIONS

The wider beams of modern MDCT scanners and wide cone beams result in increased penumbral asymmetry and heel effect, calling into question the applicability of the earlier *flat anode* model (Gagne 1989) which, additionally, only applied on the axis of rotation.

The model developed herein to describe the primary beam axial dose profile $f_p(z)$ transcends the limitations of the aforementioned model and includes the complexities of anode-tilt (asymmetric penumbra); the heel effect; the x-ray energy spectrum; and its extension to the peripheral phantom axes with these complexities still intact.

Unfortunately, inclusion of the anode-tilt produces a z-dependence of the focal spot LSF which breaks the *shift-invariant* symmetry necessary for convolution, invalidating the simple convolution [Eq. (4.11)] of the *flat anode* model of Gagne (Gagne 1989) on the AOR. However, *a method was found to recover the convolution in the form of Eq. (4.13), while still retaining the anode-tilt, without any significant compromise of accuracy.* This allowed a greatly simplified analysis of the physics theorems presented herein, as well as providing the luxury of an analytical representation for the primary beam profile $f_p(z)$ [Eq. (4.17)]. This model provides an excellent *match* to the experimental primary beam profiles (both on and off the central ray).

Additionally, a common mathematical and conceptual framework for describing the cumulative dose for both helical and axial scan series on the peripheral axes was developed by introducing a function $\bar{f}(z)$ [Eq. (4.27)], which is the helical analog of the axial, single-slice dose profile.

Using our beam model, both $f(z)$ and $\bar{f}(z)$ can be synthesized on a peripheral axis, from which the total cumulative dose resulting from either a helical and axial scan series can be readily simulated [Eq. (4.28)]. This model and simulation can provide significant insight into the nature of the quasi-periodic cumulative dose distribution on the peripheral axes. The simulated and measured dose distributions for a limited test set of helical scans were compared with generally good results. Finally, it is felt that the generality of this beam model will perhaps lead to its use in other applications beyond phantom dosimetry. In the following Chapter 5, a better scatter LSF is determined based on a Monte Carlo simulation of a "knife-edge" primary beam impulse.

The proportionality of the integral of the dose profile $f(z)$ to the product of collimator aperture a and emitted dose A_0 [Eq. (4.19)], was shown to devolve from conservation of energy, leading to the definition of *CTDI-aperture* which exhibits a fundamental constancy – its value being independent of the aperture setting (or detector configuration $nT = "N \times T"$), and which also represents the lower limit for the conventional CTDI value in MDCT.

ACKNOWLEDGMENTS

We would like to thank Richard H. Hogan and Jeffrey Cartwright of our Radiology Service Department for their assistance in obtaining the experimental data, and Thomas L. Toth, Principal Engineer – Global CT Systems, GE Medical Systems, Milwaukee, WI, for providing the aperture data and other information.

APPENDIX A: DETAILS OF PRIMARY BEAM MODEL ON THE AXIS OF ROTATION

Beginning where the text left off with the full-blown Eq. (4.3) for the primary beam axial dose (or dose rate) profile, if the small angle approximation is made (setting $\cos\varphi = 1$), such that the phantom and beam filter attenuation factors can be removed from the integral over the source coordinates and put into the energy integral, with the beam filter term being absorbed into the spectrum such that $\tilde{S}_0 E) = S_0(E)\exp[-\mu_f(E)d_f]$ now represents the photon spectrum emerging from the gantry beam port along the central ray; then the primary beam axial dose profile can be written,

$$\dot{f}_p(z) \cong \int_E \dot{A}_0(E)dE \int_s \frac{1}{\ell}g\left(\frac{s}{\ell}\right)e^{-\mu(E)\Delta d(\varphi)}\left(1 - \frac{s}{F}\cos\alpha\right)^{-2}\Pi\left(\frac{z_c}{w}\right)ds \qquad (A.1)$$

where the energy integrand has been combined into,

$$A_0(E) = \frac{\kappa(E)\tilde{S}_0(E)e^{-\mu_p(E)R}}{4\pi F^2} = \left\{\begin{array}{l}\text{emitted dose per unit energy}\\\text{at the central phantom axis}\end{array}\right\} \qquad (A.2)$$

A.1 The "Inverse Square" Correction Term

$$\left(1 - \frac{s}{F}\cos\alpha\right)^{-2} \tag{A.3}$$

results from the fact that part of the focal spot with $s > 0$, is closer to the axis of rotation than the other half at negative s. Inspection of Figure 4.1, shows that this can only affect the beam profile in the penumbra region, since each point z in the central region receives photons from the entire focal spot (both sides $\pm s$). This term does produce a small excursion in the penumbra regions at $z = \pm a/2$ of $\pm 0.8\%$, respectively, which would be barely noticeable in these high gradient regions. It will therefore be ignored.

A.2 Heel Effect Term

It can be demonstrated from Figure 4.2 without approximation that the path length in the anode is given by,

$$d(\varphi) = d_0 \frac{\sin\alpha}{\sin(\alpha - \varphi)} \tag{A.4}$$

where $\tan\alpha = z_\alpha/F$, and $\tan\varphi$ can be computed using Eq. (4), with $z' = s\sin\alpha$, as,

$$\tan\varphi = \frac{z - z'}{F(1 - z'/z_\alpha)} \tag{A.5}$$

The heel effect variation in Eq. (4.3) is $e^{-\mu(E)\Delta d(\varphi)}$, where $\Delta d(\varphi) = d(\varphi) - d_0$.

An empirical fit to the observed heel effect slope predicts an average value of $\mu d_0 = 0.28$.

The difference in path length $d(\varphi)$ for each end of the source at $s = \pm\ell/2$ compared to its center at $s = 0$ is small enough such that $\exp[-\mu\Delta d(\varphi)]$ differs from its central value by $\leq 0.2\%$, and thus, can be approximated by the center ("point source") value $\exp[-\mu\Delta d(\varphi_0)]$, where $\tan\varphi_0 = z/F$, which is independent of the source coordinate s and depends only on z. Therefore, $e^{-\mu(E)\Delta d(\varphi)}$ can be removed from the integral over the source coordinate s (or z') in Eq. (4.3) into the energy integral.

Using the small angle approximation, $\sin\varphi \cong \tan\varphi$, and $\cos\varphi \cong 1$, it can be readily shown from Eqs (A.4 and A.5) that,

$$\Delta d(\varphi_0) = \Delta d(z) = d_0 \frac{z/z_\alpha}{1 - z/z_\alpha} \cong d_0 \frac{z}{z_\alpha}\left(1 + \frac{z}{z_\alpha} + \cdots\right) \tag{A.6}$$

and it is also evident from the magnitude of μd_0 that the exponential can be approximated by,

$$e^{-\mu(E)\Delta d(\varphi_0)} \cong [1 - \mu(E)\Delta d(z)] \approx 1 - \mu(E)d_0\left(\frac{z}{z_\alpha}\right)\left(1 + \frac{z}{z_\alpha}\right) \tag{A.7}$$

In fact, the results of the approximation in Eq. (A.7) agree with the value of the actual function $\exp(-\mu\Delta d)$ using the exact value for Δd derived from Eq. (A.4), to better than 0.16%.

The integral over the energy spectrum becomes,

$$\int_E A_0(E)e^{-\mu(E)\Delta d(z)}dE \cong \int_E A_0(E)[1-\mu(E)\Delta d(z)]dE = A_0\left\{1-\langle\mu\rangle\Delta d(z)\right\} \equiv \rho(z)A_0 \quad (A.8)$$

where $\langle\mu\rangle$ is the average of $\mu(E)$ over the spectrum $A_0(E)$, and,

$$A_0 = \int_E A_0(E)dE = \begin{cases} \text{total "emitted dose" on the AOR, i.e., the} \\ \text{maximum dose possible on the axis at } z=0 \end{cases} \quad (A.9)$$

with $\rho(z)$ well-approximated by,

$$\rho(z) \cong \left[1-\langle\mu\rangle d_0\,\frac{z}{z_\alpha}\left(1+\frac{z}{z_\alpha}\right)\right] \quad (A.10)$$

Thence the complex Eq. (4.3) is reduced to the much simpler form,

$$f_p(z) \cong \rho(z)A_0\int_s \frac{1}{\ell}g\left(\frac{s}{\ell}\right)\Pi\left(\frac{z_c}{w}\right)ds \equiv \rho(z)\tilde{f}_p(z) \quad (A.11)$$

where $\tilde{f}_p(z)$ is defined as the primary dose profile with anode-tilt only, with no heel effect.

A.3 Uniformity of Phantom Attenuation Across the Slice

The assumption of uniform phantom attenuation is justified in more detail by deriving an actual "cone beam" correction term as follows.

If we are not so hasty in removing the phantom attenuation factor from the source integral, but retain a term $\exp[-\mu_p(E)\Delta R(\varphi)]$ therein due to photons at the edge of the slice obliquely penetrating an extra thickness of phantom $\Delta R = R[(\cos\varphi)^{-1}-1]$, and follow the previous heel effect derivation method, it likewise emerges that a point source approximation is applicable such that,

$$\cos\varphi \approx \cos\varphi_0 = \frac{F}{\sqrt{F^2+z^2}} = \frac{1}{\sqrt{1+z^2/F^2}} \quad (A.12)$$

$$\Delta R = R[(\cos\varphi)^{-1}-1] = R\left[\sqrt{1+z^2/F^2}-1\right] \approx R\left(\frac{z^2}{2F^2}\right) \quad (A.13)$$

Since $\Delta R = \Delta R(z)$, the exponential can be moved out of the source integral into the energy integral. Expanding the exponential, and re-evaluating the energy integral, leads to a new correction to the amplitude where $\rho(z)$ is replaced by,

$$\tilde{\rho}(z) = \rho(z)+\delta\rho(z) \approx \rho(z)-\langle\mu_p\rangle R\frac{z^2}{2F^2} \quad (A.14)$$

where $\langle \mu_p \rangle$ is the weighted average of $\mu_p(E)$ over the spectrum of photons $A_0(E)$ at the axis (i.e., the spectrum of primary photons actually reaching the axis). This will therefore represent a harder beam than the incident beam.

A reasonable value (Tsai et al. 2003) to derive a "phantom-air ratio" is. $\langle \mu_p \rangle \approx 0.224 \, \mathrm{cm}^{-1}$. Using this value and the maximum value of $R = 16$ cm, $\delta\rho(z) \leq 0.0006$ (0.06%) for a 20 mm beam width, and $\delta\rho(z) \leq 0.0024$ (0.24%) for a 40 mm beam width, both of which are negligible.

Thus, the phantom appears "transparent" and the relative beam shape on the axis of rotation "in-phantom" will be essentially the same as that "in air" at isocenter.

For even wider beam widths, one can carry the correction factor $\delta\rho(z)$ which will produce a small, symmetric "droop" at the shoulders of the profile.

A.4 Approximation of the Constraint Equation

It is convenient (although not necessary) to change the integration variable s and focal spot length ℓ to their optical projections perpendicular to the beam central ray,

$$z' = s \sin \alpha \tag{A.15}$$

$$c_0' = \ell \sin \alpha = \text{conventional definition of focal spot length} \tag{A.16}$$

from which the following scaling is also noted, $(z' / c_0') = (s / \ell)$, thence $g(z' / c_0') = g(s / \ell)$, preserving its functional form.

The primary beam profile can then be written,

$$f_p(z) \cong \rho(z) A_0 \int_s \frac{1}{c_0'} g\left(\frac{z'}{c_0'}\right) \Pi\left(\frac{z_c}{w}\right) dz' \tag{A.17}$$

A convenient scaling parameter is $z_\alpha = F \tan \alpha$, where z_α is the value of positive z beyond which all x-rays are "cut off" by the bevel of the anode surface, and it also follows that,

$$\frac{s}{F} \cos \alpha = \frac{z'}{z_\alpha} \tag{A.18}$$

It is also convenient to replace the collimator aperture w with its projected value on the axis of rotation (z-axis), $a = Mw$, where $M = (F/F_c)$.

With this change of variables, $s \rightarrow z'$, the constraint equation, Eq. (4.4), becomes,

$$\frac{z_c}{w} = \frac{z[1 - M(z'/z_\alpha)] + (M-1)z'}{a[1 - (z'/z_\alpha)]} \cong \frac{z}{a} + (M-1)\frac{z'}{a}\left(1 - \frac{z}{z_\alpha}\right) \tag{A.19}$$

where the approximation indicated in Eq. (A.19) (made by expanding the denominator and ignoring terms in z'/z_α compared to z/z_α) is accurate to 0.04%.

A.5 Solving the Tilted Anode Problem for $\tilde{f}_p(z)$

It only remains then to perform the source integration in Eq. (A.17), utilizing the relationship between the variables z' and z_c given in Eq. (A.19).

Defining the parameters $\bar{g}(z) = g(-z)$, $c(z) = (1 - z/z_\alpha)c_0$, from which it follows that $c'(z) = (1 - z/z_\alpha)c_0'$; and changing the integration variable from z' to the projection of the collimator plane coordinate z_c onto the AOR $\xi = Mz_c$, applying the constraint Eq. (A.19), we have,

$$\xi = Mz_c = \frac{a}{w}z_c = z + \left(1 - \frac{z}{z_\alpha}\right)(M-1)z' \qquad (A.20)$$

Noting that the scaling $z'/c_0' = (\xi - z)/c(z)$ applies, we have,

$$\tilde{f}_p(z) = A_0 \int_\xi \frac{1}{c(z)} \bar{g}\left(\frac{z-\xi}{c(z)}\right)\Pi\left(\frac{\xi}{a}\right)d\xi \qquad (A.21)$$

This integral may at first glance appear to have the form of a convolution, but closer inspection reveals that it does not because of the z-dependence of $c(z)$.

The significance of $c(z) = (M-1)c'(z)$ can be understood as follows (referring to Figure 4.2). If the tilted focal spot were "viewed" from a point z on the AOR offset from the central ray by an angle φ_0 ($z = F\tan\varphi_0$), its "optical projection" (or "apparent length") would be $\ell\sin(\alpha - \varphi_0)$, which can be shown to equal,

$$\ell\sin(\alpha - \varphi_0) = c_0'(1 - z/z_\alpha) = c'(z) \qquad (A.22)$$

thus $c'(z)$ is the "effective focal spot length" as seen from varying locations along the anode–cathode axis (z-axis), as one can readily picture from Figure 4.1. Then $c(z) = (M-1)c'(z)$ is the penumbra projected by $c'(z)$ onto the AOR .

We now return to the main body of the paper for further discussion and evaluation of Eq. (A.21).

APPENDIX B: PERIPHERAL PHANTOM AXES

As seen from Figure 4.5, the transformation from the focal spot coordinate system FS(x',y',z') to the peripheral axis coordinate system P(x,y,z) is a simple rotation by the angle ψ about the z'-axis which is described by,

$$\begin{bmatrix} x \\ y \end{bmatrix} = \begin{bmatrix} \cos\psi & \sin\psi \\ -\sin\psi & \cos\psi \end{bmatrix} \cdot \begin{bmatrix} x' \\ y' \end{bmatrix} \qquad z = z' \qquad (B.1)$$

Equations for the planes of interest are
  Focal spot plane:

$$y' = z'\cot\alpha \qquad (B.2)$$

Anode surface plane:

$$y' = d_0 + z' \cot \alpha$$

or equivalently,

$$y \cos \psi + x \sin \psi - z \cot \alpha - d_0 = 0 \tag{B.3}$$

Collimator plane:

$$y' = F_c \tag{B.4}$$

B.1 Heel Effect – Peripheral Axis

To calculate the path length in the anode material, one must calculate the intersection point $(0, y_i, z_i)$ of a ray from the origin in FS (center of focal spot) to a point at $(0, 0, z_p)$ on the axis P with the anode surface plane, the ray equation being given by,

$$\tan \varphi_0 = \frac{z_p}{F(\theta)} = \frac{z}{y} \tag{B.5}$$

It is straightforward (although tedious) to calculate the path length,

$$d(\varphi_0) = r_i = \frac{z_i}{\sin \varphi_0} = \frac{d_0}{\cos \psi} \frac{\sin \alpha}{\sin(\alpha - \varphi_0)} \tag{B.6}$$

and which is the same as our previous result for the AOR, Eq. (A.4), except that d_0 has been replaced by $(d_0 / \cos \psi)$.

This is logical, since the depth of anode penetrated along the central plane ($z_p = 0$, $z_i = 0$, $\varphi_0 = 0$) is $y_i = d_0 / \cos \psi$, which is simply the new reference depth from which to measure the heel effect variation from the point of view of P, hence,

$$\Delta d(\varphi_0) = d(\varphi_0) - \frac{d_0}{\cos \psi} \cong \frac{d_0}{\cos \psi} \left\{ \frac{F}{F(\theta)} \frac{z}{z_\alpha} \left[1 + \frac{F}{F(\theta)} \frac{z}{z_\alpha} \right] \right\} \tag{B.7}$$

which is the same result we had previously for the axis of rotation apart from the $\cos \psi$ factor and the required scaling of z_α from F to the distance of the peripheral axis $F(\theta)$.

B.2 Constraint Equation for a Peripheral Axis

As noted in the main text, we are still justified in utilizing a single line source in the center of the focal spot area to represent the focal spot intensity along z'.

From Figure 4.2, if the source coordinate is represented by a vector **s**, and the ray (photon path) from point s on the focal spot to a point on the z-axis of P, by a vector **r**, then,

$$\mathbf{r} + \mathbf{s} = F(\theta)\mathbf{j} + z\mathbf{k} \tag{B.8}$$

The source vector, transformed into the frame of reference $(\mathbf{i}, \mathbf{j}, \mathbf{k})$ of P, is,

$$\mathbf{s} = s\cos\alpha[\hat{\mathbf{i}}\sin\psi + \hat{\mathbf{j}}\cos\psi] + (s\sin\alpha)\hat{\mathbf{k}} \tag{B.9}$$

Solving (B.8) for $\mathbf{r}$, and using (B.9), gives,

$$\mathbf{r} = (-s\cos\alpha\sin\psi)\hat{\mathbf{i}} + [F(\theta) - s\cos\alpha\cos\psi]\hat{\mathbf{j}} + (z - s\sin\alpha)\hat{\mathbf{k}} \tag{B.10}$$

The intersection of $\mathbf{r}$ with the rotated collimator plane is represented by,

$$\mathbf{r}_c = (-s\cos\alpha\sin\psi)\hat{\mathbf{i}} + [F_c(\theta) - s\cos\alpha\cos\psi]\hat{\mathbf{j}} + (z_c - s\sin\alpha)\hat{\mathbf{k}} \tag{B.11}$$

where $F_c(\theta) = F_c / \cos\psi$.

These vectors have small (≤ 2 mm) x-components, however, since they take on both positive and negative values following s, their net contribution is negligible, and the x-components are ignored, resulting in the constraint equation,

$$\tan\varphi = \frac{z - s\sin\alpha}{F(\theta) - s\cos\alpha\cos\psi} = \frac{z_c - s\sin\alpha}{F_c(\theta) - s\cos\alpha\cos\psi} \tag{B.12}$$

which is quite similar to our previous Eq. (4.4), except for the $\cos\psi$ factor attached to the y-components.

TABLE B.1   Model Modifications Required for the Peripheral Axes

$f_p(z,\theta) = \rho(z,\theta)\tilde{f}_p(z,\theta)$	$f_p(z,\theta) = A(\theta)\int_\xi \frac{1}{c(z,\theta)}\bar{g}\left(\frac{z-\xi}{c(z,\theta)}\right)\Pi\left(\frac{\xi}{a(\theta)}\right)d\xi$
$a(\theta) = \frac{F(\theta)}{F_c}w = \frac{F(\theta)}{F}a$	$M(\theta) = \frac{F(\theta)}{F_c(\theta)}$
$c_0(\theta) = [M(\theta) - 1]c_0'$	$c(z,\theta) = c_0(\theta)\left[1 - z/z_\alpha(\theta)\right]$
$z_\alpha(\theta) = \frac{F(\theta)\tan\alpha}{\cos\psi}$	$\frac{a(\theta)}{z_\alpha(\theta)} = \frac{a}{z_\alpha}\cos\psi$
$F(\theta) = \sqrt{F^2 + R_0{}^2 - 2FR_0\cos\theta}$	$\sin\psi = \frac{R_0\sin\theta}{F(\theta)}$
$A = \frac{1}{2\pi}\int_{-\pi}^{\pi} A(\theta)d\theta$	$\rho(z,\theta) \cong \frac{\langle\mu\rangle d_0}{\cos\psi}\left\{\frac{F}{F(\theta)}\frac{z}{z_\alpha}\left[1 + \frac{F}{F(\theta)}\frac{z}{z_\alpha}\right]\right\}$
$\delta\ell = \left[R^2 - F^2\sin^2\psi\right]^{1/2} - \ell/2$	$\ell = 2\left[R_0^2 - F^2\sin^2\psi\right]^{1/2}$
$A(\theta) = \frac{1}{4\pi F(\theta)^2}\int_E \kappa(E)\tilde{S}_0(E)e^{-\{\mu_f(E)\delta f(\theta)\}}e^{-\{\mu_p(E)[\ell(\theta) + \delta\ell(\theta)]\}}dE$	

Since the collimators are planar and the detectors are curved, the collimators are tapered slightly such that the beam width on the outer detectors is (ideally) the same as that on the axis of rotation, requiring an "ideal" taper of $w(\psi) = w/\cos\psi$, thence the aperture and penumbra project as,

$$a(\theta) = \frac{F(\theta)}{F_c}a, \quad c_0(\theta) = \left[\frac{F(\theta)}{F_c(\theta)} - 1\right]c_0' = [M(\theta) - 1]c_0' \qquad (B.13)$$

If one then follows step-by-step the previous derivation for the AOR, it is straightforward (almost self-evident) that the primary beam function on a peripheral axis is given by an equation similar to that used for the AOR, as given by Table B.1 and Eq. (4.23 or 4.24) in the main body of the paper, to which we now return in order to apply these results.

GLOSSARY OF MODEL PARAMETERS AND THEIR MAGNITUDES

BS: Bremsstrahlung
AOR: axis of rotation
τ: gantry rotation time (360° rotation) ≈ 1 sec
F: focal spot to axis of rotation distance $= 541$ mm
F_c: focal spot to collimator distance along central ray $= 162$ mm
w: collimator aperture on central ray of x-ray tube
a: $(F/F_c)w \equiv Mw =$ projected collimator aperture onto the AOR
$M = F/F_c$: magnification factor
α: anode target angle $= 7°$
s: focal spot length parameter in source plane (s, x') parallel to anode surface
z': $s\sin\alpha =$ optical projection of s perpendicular to central ray
$c_0' = \ell \sin\alpha$: focal spot length (optical projection perpendicular to central ray) ≈ 1 mm

$$c_0 = (M-1)c_0' = \left[\begin{array}{l}\textbf{penumbra width projected on central axis at } z = 0 \\ \textbf{due to focal spot length } c_0' \textbf{ (as if by a slit on central ray)}\end{array}\right]: \approx 2.3 \text{ mm}$$

$g(s/\ell) = g(z'/c_0')$: focal spot relative emission intensity
z_a: $F\tan\alpha =$ scaling parameter (anode cutoff z-value on AOR) $= 66.4$ mm
$c'(z) = (1 - z/z_a)c_0'$: apparent (optical) length of the tilted focal spot as viewed from a point z on the AOR
$c(z) = (M-1)c'(z)$: penumbra projected onto the AOR at z by $c'(z)$
$\tilde{f}_p(z)$: primary beam axial dose profile with anode-tilt but not including the heel effect
$f_p(z) = \rho(z)\tilde{f}_p(z)$: primary beam axial dose profile with heel effect included
$\rho(z)$: heel effect modulation factor

$$\tilde{\rho}(z) \approx \rho(z) - \langle\mu_p\rangle R\frac{z^2}{2F^2} = \left\{\begin{array}{l}\textbf{modified heel effect factor to include cone beam} \\ \textbf{phantom attenuation for beams } > 40\,\text{mm}\end{array}\right\}$$

$\kappa(E)$: fluence-to-dose conversion factor
$nT =$ "$N \times T$": total active detector length, or total (nominal) beam width, at the AOR

$$\text{CTDI}_\infty = \frac{1}{nT} \int_{-\infty}^{\infty} f(z)dz$$

$$\text{CTDI}_a = \frac{1}{a} \int_{-\infty}^{\infty} f(z)dz = CTDI\text{-}aperture$$

MSAD: "multi-slice average dose" (average cumulative dose about $z = 0$)

A_0: primary beam dose at $z = 0$

η: scatter-to-primary ratio

REFERENCES

Barrett H.H., and Swindell W., *Radiological Imaging: The Theory of Image Formation, Detection, and Processing*, Revised ed., Academic Press, San Diego, (1981).

Dixon R.L., A new look at CT dose measurement: Beyond CTDI. *Med Phys* 30, 1272–1280, (2003).

Dixon R.L., and Boone J.M., Analytical equations for CT dose profiles derived using a scatter kernel of Monte Carlo parentage with broad applicability to CT dosimetry problems. *Med Phys* 38, 4251–4264, (2011).

Dixon R.L., Munley T., and Bayram E., An improved analytical model for CT dose simulation with a new look at the theory of CT dose. *Med Phys* 32, 3712–3728, (2005).

Gagne R., Geometrical aspects of computed tomography: Sensitivity profile and exposure profile. *Med Phys* 16, 29–37, (1989).

Hsieh J., Investigation of the slice sensitivity profile for step-and-shoot mode multi-slice computed tomography. *Med Phys* 28, 491–500, (2004).

Kachelriess M., Knaup M., Penssel C., and Kalender W., Flying Focal Spot (FFS) in Cone-Beam CT, Records of the 2004 IEEE Medical Imaging Conference, In press, 2005. Abstract: IEEE Medical Imaging Conference program, page 208, October (2004).

Morgan H.T., and Luhta R., Beyond CTDI dose measurements for modern CT scanners. Abstract: AAPM annual meeting 2004, and private communication. *Med Phys*, (2004).

Mori S., Endo M., Nishizawa K., Tsunoo T., Aoyama T., Fujiwara H., and Murase K., Enlarged longitudinal dose profiles in cone-beam CT and the need for modified dosimetry. *Med Phys* 32, 1061–1069, (2005).

Nakonechny K.D., Fallone B.G., and Rathee S., Novel methods of measuring single scan dose profiles and cumulative dose in CT. *Med Phys* 32, 98–109, (2005).

Schardt P., Deuringer J., Freudenberger J., Hell E., Knüpfer W., Mattern D., and Schild M., New x-ray tube performance in computed tomography by introducing the rotating envelope tube technology. *Med Phys* 31, 2699–2706, (2004).

Toth T.L., Bromberg N.B., Pan T.S., Rabe J., Woloschek S.J., Li J., and Seidenschnur G.E., A dose reduction x-ray beam positioning system for high-speed multislice CT scanners. *Med Phys* 27, 2659–2668, (2000).

Tsai H.Y., and Tung C.J., Analyses and applications of single scan dose profiles in computed tomography. *Med Phys* 30, 1982–1989, (2003).

Zhu X., Yoo S., Jursinic P.A., Grimm D.F., Lopez F., Rownd J.J., and Gillin M.T., Characteristics of sensitometric curves of radiographic films. *Med Phys* 30, 912–919, (2003).

CHAPTER **5**

Cone beam CT Dosimetry

*A Unified and Self-Consistent Approach
Including All Scan Modalities – With
or Without Phantom Motion*

5.1 INTRODUCTION

CT systems having beam widths along the z-axis wide enough to cover a significant anatomical length in a single axial rotation are widely available. Some utilize a conventional CT platform and can provide both conventional helical or axial scanning motions involving patient/table translation, as well as single (or multiple) rotation acquisitions at a fixed z location without table motion (used for example in acquiring cardiac images using sub-second scans) having selectable nominal cone beam widths of 40 to 160 mm in one 320 channel system. The methodology introduced herein for stationary phantom cone beam CT (SCBCT) *also applies to any CT scan without table motion* (whether wide cone beam or narrow fan beam) in applications ranging from CT fluoroscopy, perfusion studies, and multi-phasic liver scans.

A primary objective of this paper is the description of a self-consistent methodology which can bridge the gap between the dose accrued in conventional helical or axial scan modes in which the phantom is translated through a distance L; and CT operation without table/phantom motion, usually (but not necessarily) utilizing wider cone beams having variable lengths of 40–180 mm along z.

The experimental data (Mori et al. 2005) obtained on a 256 channel cone beam CT scanner (the prototype of the above-mentioned 320 channel scanner) are used to corroborate the theory and conclusions. This system and data set is also representative of the commercially available 320 channel system – the basic principles are unchanged.

For brevity, the two modalities are referred to as:

1. *"Conventional CT"*: Axial or helical scan acquisitions using multiple rotations, regularly spaced along z due to table/phantom translation over $(-L/2, L/2)$.

2. *"Stationary Phantom Cone Beam CT (SCBCT)"*: Image data is acquired using single or multiple axial rotations about a stationary phantom (table advance $b=0$, scan length $L=0$).

(The results are also applicable to narrow "fan beam" CT using a stationary phantom as used in perfusion studies.)

It will be shown in Section 5.2.3 that the dose on the central ray of the cone beam $f(0)$ is both *spatially co-located and numerically equal* to the dose predicted by CTDI for a conventional scan series for the same directly irradiated length of the phantom and thus $f(0)$ is the logical (and unique) choice for a SCBCT dose-descriptor consistent with the CTDI-based dose used in conventional CT. In addition to a common mathematical formalism which describes the dose for both modalities, there is an identical measurement technique applicable to both cases utilizing a short ionization chamber (Dixon and Ballard 2007) as illustrated in Chapter 3. It is this methodology which is recommended in AAPM Report 111 (AAPM 2010).

Both modalities are shown to possess a common equilibrium dose parameter A_{eq} which is independent of z-collimator aperture a (or $N \times T$), and a common analytical function $H(\lambda)$ is derived describing the relative approach to scatter equilibrium at $z=0$ (with $\lambda = a$ for the stationary phantom or $\lambda =$ scan length L). This commonality provides a crossover or bridge between conventional and stationary phantom CT, such that one can predict the complete data set for both modalities from a single measurement of the central (peak) dose $f(0)$ resulting from a single axial rotation at a given aperture setting a. From this, one can predict the SCBCT or fan beam peak dose $f(0)$ for any beam width a, and the conventional CT dose (as predicted by CTDI) for any scan length L (including the limiting equilibrium dose) for any collimator aperture setting a and any pitch p. Although the crossover between modalities is an interesting aspect of the theory developed, it is by no means the only application or goal of the theoretical development to follow.

The glossary of parameters below is provided as a quick reference for the following development and throughout.

The following development concentrates on the *actual in-phantom dose* for both conventional CT involving phantom translation, and stationary phantom CT such as SCBCT, directed toward creating a consistent approach to CT dose assessment which provides continuity of dose and physical interpretation between these two modalities. To that end, it is necessary to review the dose-descriptors used in conventional CT (and the CTDI-paradigm).

5.2 THEORY

5.2.1 Conventional CT Scanning Using Table/Phantom Translation: Accumulated Dose Equations for Helical or Axial Scan Trajectories Utilizing Table/Phantom Translation Along z

As shown in Chapter 2, the equation for the accumulated dose for helical or axial scanning is given by,

$$D_L(z) = \frac{1}{b} f(z) \otimes \Pi(z/L) = \frac{1}{b} \int_{-L/2}^{L/2} f(z-z')dz' \qquad (5.1)$$

where $f(z)$ includes both the primary beam and scatter contributions $f(z) = f_p(z) + f_s(z)$ and is therefore much broader than the *primary beam width (fwhm) a*, where a is equal to the z-collimator aperture geometrically projected onto the axis of rotation (AOR) (Dixon et al. 2005). Multi-detector CT (MDCT) requires that $a > nT$ in order to keep the penumbra beyond the active detector length nT. Evaluation of Eq. (5.1) at $z = 0$ results in the accumulated dose $D_L(0)$ *at the center of the scan length L* as given by Eq. (5.2),

$$D_L(0) = \frac{1}{b} \int_{-L/2}^{L/2} f(z')dz' \tag{5.2}$$

which (for axial scans) represents an average dose over the small interval $\pm b/2$ about $z = 0$, where b is typically small compared to the total scan length $L = Nb$. Note, again, the implicit dependence of the integration limits $\pm L/2$ and the divisor b, physically related by $L = Nb$.

Eq. (5.2) for $D_L(0)$ represents the basic equation upon which the CTDI methodology is based, in which the divisor b of the integral physically represents a table advance per rotation, with $CTDI_L$ itself defined (Shope et al. 1981) as the value of $D_L(0)$ in Eq. (5.2) resulting from a specific table increment (scan interval) $b = nT$, the interval of which produces "contiguous" axial scans in the image domain (leaving no gaps in the *acquired image data*). Thus, physically $CTDI_L$ is equal to the *accumulated dose* at the center ($z = 0$) of the scan length $(-L/2, L/2)$, for a table advance $b = nT =$ "N×T" (a generalized pitch $p = b/nT = 1$). Substituting $b = nT$ into Eq. (5.2) gives the familiar CTDI equation,

$$CTDI_L = \frac{1}{nT} \int_{-L/2}^{L/2} f(z')dz' \tag{5.3}$$

which one can also express in terms of $D_L(0)$ from Eq. (5.2) and pitch p as $CTDI_L = (b/nT)D_L(0) = pD_L(0)$.

There is actually no imperative to have a separate equation for $CTDI_L$, since it simply represents a special case of Eq. (5.2), namely, a particular value of $D_L(0)$ corresponding to a specific table increment $b = nT$ (a pitch of unity), and thus nT in the CTDI equation *physically* represents a table increment.

Measurement of the integral in Eqs (5.2 or 5.3) using a pencil chamber of fixed length ℓ only allows prediction of the accrued dose at $z = 0$ for a scan length $L = \ell$ (e.g., $L = 100$ mm as for $CTDI_{100}$). Additionally, assigning a fixed integration length L (e.g., 100 mm) to $CTDI_L$ breaks the required coupling $L = Nb$ between the divisor b and the integration limits $\pm L/2$. For conventional CT, as L becomes large enough to completely span the very long scatter tails of $f(z)$ at $L = L_{eq}$, such that no additional scatter can reach $z = 0$ for $L \geq L_{eq}$ (symbolically $L \to \infty$), then $D_L(0)$ approaches its limiting value – the equilibrium dose D_{eq}, written as,

$$D_{eq}(a/b) = \frac{1}{b} \int_{-\infty}^{\infty} f(z')dz' \propto (a/b). \tag{5.4}$$

Since D_{eq} depends (explicitly) on the inverse of table increment b, and is directly proportional to the collimator aperture a implicitly through the infinite integral of $f(z)$ (as shown in Chapter 3), then D_{eq} is directly proportional to a/b.

5.2.2 Helical Scanning

The derivation of Eqs (5.1–5.4) for helical scanning in Chapter 2 is briefly outlined in this section for transition. The dose rate on the *phantom central axis* is independent of beam (gantry) angle θ, hence the *dose rate* profile is $\dot{f}(z) = \tau^{-1} f(z)$, where τ is gantry rotation time. Translation of the table and phantom at velocity v produces a dose rate profile in the phantom reference frame expressed as a traveling wave $\dot{f}(z,t) = \tau^{-1} f(z - vt)$, thus the dose accumulated at a given z as the profile travels by is given by the time-integral of $\dot{f}(z,t)$ over the total "beam-on" time t_0, namely,

$$D_L(z) = \tau^{-1} \int_{-t_0/2}^{t_0/2} f(z - vt)dt.$$

Conversion to the spatial domain using $z' = vt$, scan length $L = vt_0$, and a table advance per rotation $b = v\tau$ (a pitch of $p = b/nT$), leads directly to the convolution equation for $D_L(z)$ [Eq. (5.1)] from which Eqs (5.2–5.4) also follow as before; however, $D_L(0)$ and $CTDI_L$ in *the helical mode* both refer to the dose *precisely* at $z = 0$ (and likewise on the *peripheral axis* where an angular average (Dixon 2003) is used as illustrated in Chapter 2).

5.2.2.1 Transition from Helical to Stationary Table/Phantom

If the table and phantom remain stationary ($v = 0$), the time-integral of the dose rate $\dot{f}(z) = \tau^{-1} f(z)$ is simply $D_L(z) = (t_0 / \tau)f(z) = Nf(z)$; likewise, in the limit $v \to 0$, $L = vt_0 \to 0$, the integration limits $\pm L/2 \to 0$; thus the integral format of Eqs (5.1–5.3) collapse, and all converge smoothly to $Nf(z)$ or $Nf(0)$. This convergence is readily seen using Eq. (5.2). As L becomes very small, the integral can be approximated by $f(0)L$, thence $D_L(0) \approx (1/b)f(0)L = Nf(0)$, or formally as $\lim_{L \to 0} D_L(0) = Nf(0)$.

The quantities b and $L = Nb$ are *dynamic variables of table/phantom motion* (and intimately coupled by this motion), therefore artificially constraining one or both in Eqs (5.1–5.3) (such as fixing L in $CTDI_{100}$) will foil the convergence to $Nf(0)$ as $b \to 0$, and thus negate its relevance to SCBCT. Neither $Nf(z)$ nor $Nf(0)$ contain nT which likewise has no relevance in SCBCT dosimetry (in which scan "contiguity" and pitch play no role).

5.2.2.2 The Following Important Points Are Clear From the Foregoing:

1. The integral of $f(z)$ over $(-L/2, L/2)$ for $D_L(0)$ in Eq. (5.2) and for $CTDI_L$ in Eq. (5.3) *is solely the result of phantom translation* over the distance $L = vt_0 = Nb$; moreover, the integration limits $\pm L/2$ and the divisor of the integral (the table increment $b = v\tau$) are necessarily coupled via $L = vt_0 = Nb$ (coupled by couch velocity v for helical scans).

2. The integral of $f(z)$ over $(-L/2, L/2)$ does not imply any averaging of the dose over the scan length L, but rather $CTDI_L$ predicts the dose precisely at $z = 0$ at the center of the scan length $(-L/2, L/2)$ for helical scans at a pitch of unity.

3. Physically, this integral represents a summation of the decreasing incremental contribution to the dose at $z = 0$ by the scatter tails of the traveling profile as it gets further from the origin; and it is this lateral dispersal of the dose profiles due to phantom translation which results in the central dose at $z = 0$ reaching a limiting equilibrium value D_{eq} for large scan lengths.

4. CTDI always predicts the dose at (or about) the center of the scan interval $(-L/2, L/2)$ at $z = 0$.

5. For a stationary phantom ($v \to 0$, and $b = v\tau \to 0$, $L = vt_0 \to 0$), and since the integration length $L \to 0$, the integral format of Eqs (5.1–5.3) "collapses" – smoothly converging in this limit to the non-integral form $D_L(z) = Nf(z)$, or $D_L(0) = Nf(0)$, which increase without bound with N, since the individual dose profiles (deprived of lateral dispersal due to phantom translation) simply pile up on top of each other.

5.2.2.3 In Summary

For conventional CT scanning, $CTDI_L$ (or any dose $D_L(0) = p^{-1}CTDI_L$ derived from it) *always* represents the accumulated dose at (or about) the center of the scan length $(-L/2, L/2)$ at $z = 0$. This also applies to MSAD (Shope et al. 1981), $CTDIw$ (Leitz et al. 1995) and $CTDI_{vol}$ – the latter two are essentially *planar averages* over the *area* of the central scan plane at $z = 0$, since no averaging over the scan length L has been performed.

5.2.3 The Case of the Stationary Phantom

The dose distribution produced by a single axial rotation of a wide cone beam (or a narrow fan beam) having a primary beam width $(fwhm) = a$ (where a is also the projected z-collimator aperture setting) is denoted by $f(z)$ as before, and by $Nf(z)$ for N rotations without table translation, with a central ray peak dose $Nf(0)$. It is simplicity itself compared to scans with table motion and is easy to simulate and quantify. No integral is involved as with the CTDI-paradigm and *no pencil chamber is required (or desired)*.

5.2.3.1 Relating the Dose and the Dose Distribution in SCBCT to That of Conventional CT

- The fact that $CTDI_L$ represents the *dose* at the center ($z = 0$) of the scan length $(-L/2, L/2)$ in conventional CT suggests that its direct analogue in the case of SCBCT would likewise be the *dose* $Nf(0)$ at the center $z = 0$ of the beam $(-a/2, a/2)$ – viz., on the "central ray" of the cone beam; corresponding to the location of the maximum or "peak" dose for both modalities.

- This logical conclusion is further supported by the physics which suggests that there should be little difference in the dose distribution $f(z)$ produced by a wide primary beam of width a (a cone beam) and the axial dose distribution $\tilde{D}_N(z)$ produced by a series of N adjacent, narrow primary beams of width $\hat{a} = a/N$, spaced at intervals $b = \hat{a} = a/N$, resulting in the same total energy deposition in the phantom as the cone beam, while "directly irradiating" (with the primary beam) the same length of phantom $L = Nb = a$ in both cases. Likewise, for helical scanning at pitch $p = \hat{a}/nT$.

• This correspondence can be directly confirmed using the experimental data (Mori et al. 2005) from a 256 channel cone beam CT system for which dose profiles for all available cone beam widths (apertures) a ranging from 28 mm up to $a = 138$ mm were measured. The widest cone beam $a = 138$ mm should produce the same dose distribution $f(z)$ as $N = 5$ axial profiles with $\hat{a} = 28$ mm spaced at intervals $b = \hat{a} = 28$ mm, thus covering a scan length $L = Nb = 140$ mm – a length essentially equal to the $a = 138$ mm cone beam width $(L = a)$ as depicted in Figure 5.1 which shows that the simulated axial dose distribution $\tilde{D}_N(z)(\bullet)$ resulting from the superposition (summation) of

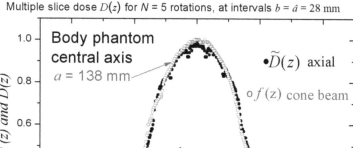

FIGURE 5.1 The axial dose profile $f(z)$ for a wide cone beam of width $a = 138$ mm generated by a single rotation about a stationary phantom (o) exhibits little difference from the accumulated dose distribution $\tilde{D}_N(z)$ (●) due to the superposition of $N = 5$ axial profiles $\hat{f}(z)$ having $\hat{a} = 28$ mm† spaced at intervals (a table increment of) $b = \hat{a} = 28$ mm, which results in a scan length. $L = N\hat{a} = 140$ mm; giving essentially the same directly irradiated length $(L = a)$ as the cone beam having $a = 138$ mm. The peak doses at $z = 0$ are essentially equal, $\tilde{D}_N(0) \approx f(0) = 1.0$, thereby corroborating the choice of $f(0)$ to represent the dose for SCBCT.

† **Rounded up from 27.5 mm thus $L = 5 \times 27.5 = 138$ mm exactly.**

(From Dixon and Boone, *Medical Physics*, 2010.)

the $N=5$ narrow axial profiles $\hat{f}(z)$ of primary width (aperture) $\hat{a}=28$ mm, laterally displaced at intervals $b=\hat{a}=28$ mm over $L=Nb=140$ mm, is seen to be essentially coincident with the cone beam dose distribution $f(z)$ of width $a=138$ mm ($L=a$) as was postulated as depicted by open circles (**o**).

Since $b=\hat{a}$ produces a smooth axial dose distribution (Dixon et al. 2005) on the AOR, then $D_L(0) \approx f(0)$ for the cone beam as is likewise evident from Figure 5.1, where $D_L(0)$ is the conventional CT dose at $z=0$ obtained from Eq. (5.2) using $b=28$ mm and $L=Nb=140$ mm (or as calculated from $CTDI_L$ using $D_L(0)=p^{-1}CTDI_L$ with $p=\hat{a}/nT$).

- This provides a direct physical connection between the dose $D_L(0)$ in conventional CT (described using $CTDI_L$) and the corresponding dose $f(0)$ in SCBCT, these doses being not only spatially co-located at $z=0$ but also equal in magnitude $D_L(0) \approx f(0)$–representing equal peak doses at $z=0$ in both cases as is clearly illustrated in Figure 5.1.

- Therefore $f(0)$ (or $Nf(0)$ for multiple rotations) is the obvious choice to represent the dose for stationary phantom CT – for wide cone beams (SCBCT) and narrow fan beams alike.

- In fact, $f(0)$ is *the only choice producing continuity between the two modalities* – use of any other "cone beam dose index" which predicted a dose other than $f(0)$ would amount *to the paradoxical assignment of different "dose values" to the same dose distribution (i.e., for the same length of anatomy imaged using the same x-ray technique)*.

- It is also satisfying to note that the mathematics automatically forces the same conclusion, with the basic equations [Eqs (5.1–5.3)], derived for conventional CT, all converging to the proper dose $Nf(z)$ for a stationary phantom in the limit as table advance $b \rightarrow 0$ as previously shown, *thereby losing their integral format* (including $CTDI_L \rightarrow Nf(0)$, since $b=nT$ is a table advance and likewise this $b \rightarrow 0$). Note that this convergence is foiled by artificially fixing the integration length of $CTDI_L$, e.g., $L=100$ mm as in $CTDI_{100}$.

5.2.3.2 Measurement of the Central Ray Dose f(0) for a Wide Cone Beam and a Stationary Phantom in SCBCT

- Since $f(0)$ is the dose on the central ray of the cone beam at depth in the phantom, the obvious (and simplest) method is to directly measure $f(0)$ at that point using a small ionization chamber (such as a 0.6cc Farmer-type chamber) – *the same method used for decades to measure depth-dose in a stationary phantom; and as recommended in AAPM Report 111 (AAPM 2010)*.

- This same measurement method (Dixon 2003; Nakonechny et al. 2005; Anderson et al. 2005; Mori et al. 2005; Dixon and Ballard 2007) has also been utilized in conventional CT to directly measure the accumulated dose $D_L(0)$ at $z=0$ during phantom translation over $(-L/2, L/2)$; its validity and robustness having been thoroughly demonstrated in Chapter 3; and which method also offers considerable advantages over the fixed-length pencil chamber – namely, unrestricted integration length (scan length) L, as well as broad applicability to shift-variant phantoms and techniques since it represents a direct dose measurement rather than a dose *inferred* from an integral acquired by irradiation of a pencil chamber.

- A pencil chamber *cannot* be used to measure $f(0)$ since it can only measure the integral of $f(z)$, *and thus cannot distinguish between dose profiles having the same area but differing "peak" (or central ray) doses $f(0)$.*

5.2.4 The Equilibrium Dose Constant A_{eq}

The Equilibrium Dose Constant A_{eq}: a useful simplification obtained by setting the table advance b equal to the aperture a ($b = a$).

Returning to conventional CT scanning with phantom translation, an interesting and useful shortcut is described. Setting the scan interval $b = a$ (where a closely approximates the primary beam *fwhm*) produces scan contiguity *in the dose domain*. Since D_{eq} is proportional to (a/b), the equilibrium dose approached when $b = a$, denoted by $A_{eq} = (b/a) D_{eq}$, will depend on neither b nor a; and [using Eq. (5.4)] A_{eq} can be written as,

$$A_{eq} = \frac{1}{a} \int_{-\infty}^{\infty} f(z')dz' = \text{constant} \tag{5.5}$$

A_{eq} is a constant, independent of aperture a (and thence nT), since the integral is directly proportional to a as previously noted. For conventional CT, $D_{eq} = (a/b)A_{eq}$, thus A_{eq} is the equilibrium dose for a table advance $b = a$ (or a pitch $p = a/nT$); and called "CTDI-aperture" in Chapter 4.

The equilibrium dose constant $A_{eq} = (a/nT)^{-1}CTDI_\infty$, although related to $CTDI_\infty$, is not equal to $CTDI_\infty$, since $b = a$ represents a detector pitch $p = a/nT > 1$ in MDCT for which a primary beam width $a > nT$ is required to keep the penumbra beyond the active detector length nT (referred to as "over-beaming"), thus $A_{eq} < CTDI_\infty$.

For conventional CT, "over-beaming" produces an increase in CTDI and accumulated dose by the factor a/nT (as compared to $a \approx nT$ for single-slice scanners); however, over-beaming has less significance in SCBCT – producing an almost trivial dose increase since there are no overlapping, *adjacent* dose profiles.

Note that the aperture a corresponding to a given nT can be calculated from the dose efficiency, which is essentially equal to the inverse of the over-beaming factor $(a/nT)^{-1}$. Most scanner technical manuals contain aperture data (or primary *fwhm* data). Typical apertures have also been published (Dixon et al. 2005; Dixon and Ballard 2007) and shown in Chapter 3 for the GE Lightspeed family of scanners. However, as noted, $(a/nT)^{-1}$ does not actually represent dose efficiency for *stationary phantom* CT (such as SCBCT).

Use of A_{eq} allows a considerable reduction in data collection as will be illustrated, since its value can be fixed by making a measurement at single value of the aperture a.

5.3 NUMERICAL ANALYSIS OF EXPERIMENTAL SCBCT DOSE DATA

The experimental data (Mori et al. 2005) obtained on a 256 channel cone beam CT scanner includes a direct measurement of both the central ray dose $f(0)$ as well as the infinite integral of the dose profile $f(z)$ denoted by DPI_∞; the data of which can be used to illustrate the magnitude by which the actual dose $f(0)$ is over-estimated should one attempt to apply $CTDI_\infty$ to the problem (or by using $CTDI_{300}$, measured using a 300 mm long pencil chamber, to approximate $CTDI_\infty$). As previously noted, nT has no relevance to the dose in stationary phantom CT (contiguity has no meaning and pitch $p = b/nT = 0$).

The analysis is both simplified and made considerably more *interesting* if one uses A_{eq} from Eq. (5.5) as a "surrogate" for $CTDI_\infty$, where $A_{eq} = (a/nT)^{-1} CTDI_\infty < CTDI_\infty$.

Table 5.1 illustrates the relationship between the measured central ray dose $f(0)$ and A_{eq} for various beam widths (aperture values) ranging from $a = 28$–138 mm, resulting from a single axial rotation about the center ($z = 0$) of a 900 mm long, 32 cm diameter stationary PMMA body phantom (Mori et al. 2005).

The original raw data (Mori et al. 2005) were re-analyzed to deduce the effective apertures a using the equivalent width (Bracewell 2000) of the *primary beam* profiles as measured free-in-air (Mori et al. 2005) [$a \approx$ width at $\frac{1}{2} f(0)$].

Thus the *equilibrium dose constant* $A_{eq} = D_{eq}(b/a)$ is indeed seen to be independent of aperture a as previously postulated, remaining constant to better than $\pm 0.7\%$ over the entire range of apertures from $a = 28$–138 mm; whereas $CTDI_\infty = (a/nT)A_{eq}$ varies by a factor of 2.5 over the same range (corresponding nT values are 10, 32, 64, 96, and 128 mm); however, the *important point is that A_{eq} significantly over-estimates the relevant SCBCT*

TABLE 5.1　Measured Data (Mori et al. 2005) – Central Axis, Body Phantom

Primary Beam Aperture a	Dose Profile Integral $DPI_\infty = \int_{-\infty}^{\infty} f(z)dz$	Equilibrium Dose Constant $A_{eq} = DPI_\infty/a$ $A_{eq} = D_{eq}(a) = \dfrac{1}{a}\int_{-\infty}^{\infty} f(z)dz$	Measured Central Ray Dose $f(0)$	Ratio $\dfrac{A_{eq}}{f(0)}$
138 mm	848 mGy. mm	6.14 mGy	4.37 mGy	1.43
111 mm	686	6.18	3.90	1.58
80 mm	498	6.22	3.19	1.95
49 mm	303	6.18	2.27	2.72
28 mm*	169	6.15	1.53	4.02

Source: Dixon and Boone, *Medical Physics* (2010).

Measured cone beam doses and dose integrals resulting from a single axial rotation about a stationary phantom (no table advance) for a 256 channel scanner (SCBCT).

Based on data of Mori et al., where $f(z)$ denotes the axial dose profile corresponding to a primary beam width a on the central axis of the PMMA body phantom (32 cm diameter, 900 mm length), integrated over 900 mm to obtain the dose profile integral DPI_∞ – all data being normalized per 100 *mAs*.

*Rounded off from 27.5 mm.

dose $f(0)$ by the factor shown in the last column of Table 5.1, viz., by a factor of 4 for a beam width of $a = 28$ mm, by a factor of 2 for $a = 80$ mm, and by a factor of 1.4 for $a = 138$ mm, with $CTDI_\infty$ giving an even larger over-estimate (since $a > nT$). This result was previously anticipated since neither A_{eq} nor $CTDI_\infty$ (described by integral equations) are relevant to the dose in stationary phantom CT (SCBCT).

5.3.1 The *Inapplicability* of the CTDI-Paradigm and the Pencil Chamber to Stationary Phantom Dosimetry

The data in Table 5.1 is also illustrative of the magnitude of the error obtained in SCBCT when using a pencil chamber of length $\ell = 300$ mm to measure $CTDI_{300}$ (as an approximation to $CTDI_\infty$) as illustrated by the following *Gedanken* experiment. Assume a cone beam width $a = 115$ mm corresponding to $nT = 100$ mm. By definition, $CTDI_{300}$ is equal to the dose at the center of *three contiguous* axial scans, each with $nT = 100$ mm, stitched together using an interval (a table advance) of $b = nT = 100$ mm, for a total scan length of $L = Nb = 300$ mm. In this case, $CTDI_{300}$ over-estimates the SCBCT peak dose $f(0)$ by about 60%, due to the inclusion of scatter from the two additional contiguous scans which augment the dose $f(0)$ at $z = 0$ [$CTDI_{300} = 1.6\,f(0)$]. The notion that the integral equations for $CTDI_\infty$ or A_{eq} might predict some useful average dose for SCBCT is readily dispelled; since it is clear from the data in Table 5.1 and Figure 5.1 that A_{eq} and thus $CTDI_\infty$ (and $CTDI_{300}$) are both larger than the peak dose $f(0)$ for every available aperture setting, i.e., *their associated "dose values" do*

not exist anywhere in the phantom. Mathematically, in order to obtain $f(0)$ from $\displaystyle\int_{-\infty}^{\infty} f(z)dz$,

one must divide the integral by the equivalent width (Bracewell 2000) a_w of the function; however, since $f(z)$ contains a broad scatter component, a_w is considerably larger than the *primary beam width a* (or nT) used as divisors in computing A_{eq} (or $CTDI_\infty$) (e.g., for the profiles having $a = 28$ mm shown in Figure 5.1, $a_w = 112$ mm, thus A_{eq} over-estimates $f(0)$ by a factor of 4). It is clear that A_{eq} (or $CTDI_\infty$) will therefore always over-estimate $f(0)$. So why do we divide the integral by the primary beam width? *We don't* – these divisors represent *table increments* $b = a$ for A_{eq} (or $b = nT$ for CTDI) and not primary beam widths (actual or nominal). *The CTDI-paradigm was never intended to predict the dose* $f(0)$ *for a single axial rotation,* but rather the accumulated dose $D_L(0)$ at $z = 0$ for N multiple scans – spaced at intervals b due to table translation over a length $L = Nb$. *A small ion chamber can precisely measure the desired peak dose* $f(0) = 1.0$ in Figure 5.1, whereas the "dose" values given by the integral "CTDI-types" are $A_{eq} = 1.4$, $CTDI_\infty = 1.5$, and $CTDI_{300} = 1.4$, *all lying well above* $f(0) = 1.0$ (even above the top of Figure 5.1).

Since A_{eq} is independent of aperture a, a narrow beam could be used to measure A_{eq} if the aperture a is known; however, this is small consolation since the values of both A_{eq} and $CTDI_\infty$ over-estimate the dose $f(0)$ in SCBCT for clinically relevant cone beam widths (see Table 5.1). The SCBCT dose $f(0)$ would exhibit a variation of more than 200% (a factor of 2.3) over the range of beam widths $a = 50$–180 mm available for clinical use on the aforementioned 320 channel SCBCT scanner (corresponding to $nT = 40$–160 mm); and a much larger variation for the narrow *"fan beams"* also used in stationary phantom CT such as in CT fluoroscopy or perfusion studies, for which the peak dose $f(0)$ may be significantly over-estimated (Bauhs et al. 2008) by attempting to apply the dose paradigm based on $CTDI_{100}$.

It is therefore important to determine $f(0)$ over the complete range of apertures used clinically in the SCBCT acquisition mode, e.g., for 40 mm $\leq nT \leq 160$ mm. A useful theoretical function is derived in Section 5.4 Eq. 5.8, describing the variation of $f(0)$ with a, which closely *matches* the data in Table 5.1; and which can be used to extrapolate a measurement of $f(0)$ at a single value of aperture a to any other aperture; thus allowing the prediction of the peak doses $f(0)$ for narrow fan beams which would require an ion chamber length $\ell < nT$ for the central axis measurement.

5.3.2 The Approach to Scatter Equilibrium for SCBCT

From the data in Table 5.1, it appears that $f(0)$ is increasing *toward* A_{eq} as the cone beam width a increases, and such a convergence does indeed occur, but only for very wide (and thus *clinically irrelevant*) cone beam widths of $a \geq 470$ mm. Like the accumulated dose $D_L(0)$ in conventional CT, the SCBCT dose $f(0)$ will also asymptotically approach a maximum equilibrium value $f_{eq}(0)$ when the cone beam width a becomes wide enough to achieve scatter equilibrium on the central ray at $z = 0$, such that scatter produced from any further increases in primary beam width can no longer reach $z = 0$, and thus no longer affect $f(0)$. Attainment of equilibrium at $z = 0$ depends only on the *distance* of the outermost primary-beam photons from the origin. When the probability becomes negligible that primary photons scattered at the far-flung primary beam edges (at $z' = \pm a/2$ for the cone beam or at $z' = \pm L/2$ in conventional CT) can reach the origin, then equilibrium is achieved for cone beam widths $a \geq a_{eq}$, and likewise for scan lengths $L \geq L_{eq}$ for conventional CT, *thus it follows that $a_{eq} = L_{eq}$. Likewise, the magnitude of the equilibrium dose constant A_{eq} is the same for both modalities.* (These results follow from our previous arguments demonstrating the equality of the dose at $z = 0$ for the two modalities for the case $L = a$).

Since $L_{eq} \approx 470$ mm in conventional CT, scatter equilibrium at $z = 0$ will therefore occur for cone beam widths $a \geq 470$ mm in the 32 cm PMMA body phantom. However, since such wide cone beams are not utilized (nor likely to be) in this modality, the cone beam equilibrium dose $f_{eq}(0) = A_{eq}$ is not clinically relevant for SCBCT; whereas, for conventional CT scanning, typical body scan lengths of $L \geq 250$ mm produce doses which closely approach D_{eq}; therefore D_{eq} or $CTDI_{\infty}$ [and likewise $A_{eq} = (b/a) D_{eq}$] is a considerably more relevant dose for this modality. However, the value of A_{eq} can serve as a convenient, common normalization constant – *for both modalities.*

5.3.3 The Approach to Equilibrium Function $H(\lambda)$

The variation of the relative approach to equilibrium function $H(a) = [f(0) / f_{eq}(0)]$ is conceptually quite similar to the increase in dose at a given depth with increasing field size observed for a stationary x-ray beam incident on a phantom (except only one field dimension a is varied in this case). In fact, the SCBCT equilibrium dose is the same as A_{eq} for conventional CT, viz. $f_{eq}(0) = A_{eq}$; and the relative approach to equilibrium curves

$H(L) = D_L(0)/D_{eq}$ and $H(a) = f(0)/A_{eq}$ should be the same using the correspondence $a = L$, based on our previous analogy between a cone beam and a juxtaposition of adjacent narrow fan beams.

5.4 MODELING THE CONE BEAM

For definiteness and simplicity, all numerical examples and derivations refer to the dose on the central axis of the 32 cm diameter cylindrical PMMA body phantom at 120 kVp using a bow-tie filter unless otherwise noted. The peripheral axis is dealt with in Section 5.4.4.

5.4.1 General Considerations

As previously seen in Figure 5.1, there is nothing particularly mysterious about the SCBCT axial dose profile $f(z)$ for a wide cone beam of width a; being quite similar to the cumulative dose distribution $D_L(z)$ in conventional CT at a pitch near unity ($p = \hat{a}/nT$) for the same directly irradiated phantom length $L = a$. Differences in beam divergence between the narrow beams and wide cone beam are quite small (only about $\pm 7°$ off the central ray at the extreme edges $z = \pm a/2$ of the widest cone beam, $a = 138$ mm).

5.4.2 The Heel Effect

The only real difference is that the wider collimator aperture of the cone beam enhances the heel effect, however, neither its odd nor its even components have any significant effect on the central ray dose $f(0)$ (the SCBCT dose-descriptor of interest), with both the primary and the scatter components at $z = 0$ being essentially unaffected. The primary beam heel effect is largely masked by the scatter component. Thus, a simple model which ignores anode-tilt and the heel effect should do quite nicely for predicting the relevant SCBCT dose $f(0)$, but may do somewhat less well in reproducing the entire dose distribution $f(z)$ – particularly near the beam edges for wide cone beams on the peripheral axes. It also produces the same results as a more complex model illustrated in Chapter 4 (Dixon et al. 2005) (which includes anode-tilt and the heel effect) for the integral theorems involving A_{eq} and D_{eq}.

5.4.3 A Simple Beam Model Predicting the Observed Dose Data

The *primary beam LSF* is the focal spot emission intensity (Dixon et al. 2005) (as slit-projected by the z-collimators onto the axis of rotation AOR), and expressed as a scaled function $lsf(z) = c^{-1}g(-z/c)$ having unit area, where c represents the slit-projected focal spot length ($c \sim 3$ mm) (see Chapter 4).

The *scatter LSF* $(z - z')$ is the much broader scatter response function to a unit-strength primary beam impulse $\delta(z')$ (or "knife edge") applied at $z = z'$, where η is the scatter to primary ratio S/P, where η is equal to the ratio of scattered to primary energy deposited along z as expressed by the ratio $\eta = S/P$, where $S = \int\limits_{-\infty}^{\infty} f_s(z')dz'$ and $P = \int\limits_{-\infty}^{\infty} f_p(z')dz'$. Thus $\int\limits_{-\infty}^{\infty} LSF(z)dz = \eta$. It is convenient to express $LSF(z) = \eta lsf(z)$ where $lsf(z)$ is a unit-area, scaled function, symbolically represented as $lsf(z) = d^{-1}h(z/d)$, where d represents the width of the broad scatter LSF ($d \sim 100$ mm) and where $d \gg c$.

The model generates the primary beam component $f_p(z)$ and the scatter component $f_s(z)$ of the axial dose profile $f(z) = f_p(z) + f_s(z)$ by the convolution shown in Eq. (5.6),

$$f(z) = \left[\frac{1}{c} g\left(\frac{-z}{c} \right) + \eta \frac{1}{d} h\left(\frac{z}{d} \right) \right] \otimes A_0 \Pi\left(\frac{z}{a} \right) \qquad (5.6)$$

where $A_0\Pi(z/a)$ represents the core primary beam function (without penumbra). The convolution of the focal spot lsf $[c^{-1}g(-z/c)]$ with $A_0\Pi(z/a)$ produces the primary beam function $f_p(z)$ by adding a penumbra of width c to $\Pi(z/a)$, and the convolution of the scatter $LSF = \eta lsf(z) = \eta d^{-1}h(z/d)$ with the primary beam core $A_0\Pi(z/a)$ gives the scatter component $f_s(z)$, where the negligible effect of the primary beam penumbra on the scatter distribution has been ignored (its effect is truly nil).

Since the penumbra c is typically small compared to the aperture setting a in MDCT, the width c of the penumbra added to $\Pi(z/a)$ by its convolution with $c^{-1}g(-z/c)$ is small compared to a ($c \ll a$), in which case as the aperture a is decreased to a value comparable to the penumbra c, the beam *fwhm* actually increases, becoming larger than a and its peak intensity $f_p(0)$ decreases below the emitted intensity A_0 due to narrow slit effects (Dixon et al. 2005). This effect required post-patient collimation to achieve narrow slice widths for single-slice scanners.

$A_0 = f_p(0)$ which is the "point dose" on the central ray ($z = 0$) contributed by the primary beam on the AOR (at a depth of 16 cm in the body phantom).

5.4.3.1 The Integral Theorem

The expression for the infinite integral of $f(z)$, and thence the equilibrium dose constant A_{eq} in Eq. (5.5) (or $CTDI_\infty$), immediately follows from Eq. (5.6) without requiring any detailed knowledge of the functional form of $f(z)$ (or that of either lsf); the infinite integral of the convolution in Eq. (5.6) being obtained by inspection as,

$$\int_{-\infty}^{\infty} f(z')dz' = A_0(1+\eta)a = f_p(0)(1+\eta)a \qquad (5.7)$$

which shows the important proportionality of the infinite integral of $f(z)$ to the aperture a, and also providing the theoretical formula for A_{eq} shown in Eq. (5.8),

$$A_{eq} = \frac{1}{a} \int_{-\infty}^{\infty} f(z')dz' = A_0(1+\eta) = f_p(0)(1+\eta) \qquad (5.8)$$

Thereby confirming the constancy of A_{eq} – namely *its independence of aperture setting a (and also of nT)*. Eq. (5.8) shows that A_{eq} depends only on the primary beam intensity $f_p(0)$ on the central ray, which is well-known to be independent of collimator setting (assuming $a > c$), and on the S/P ratio η (and since η is also the impulse-response amplitude, it cannot depend on a); and *also confirms that A_{eq} is the same for conventional CT and SCBCT.*

The equilibrium dose for conventional CT can be written as $D_{eq} = (a/b) A_{eq}$ which is directly proportional to (a/b) since A_{eq} is a constant (physically, opening the collimator aperture a deposits more energy per rotation, and reducing b packs the dose profiles into a smaller length, both leading to an increase in dose). Thus, D_{eq} for conventional CT can have different values, depending on both the aperture a (and thence nT) and the table increment b (or helical pitch $p = b/nT$), and $D_{eq} = A_{eq}$ only for a table advance of $b = a$; however, for SCBCT the scatter-equilibrium limit for the dose $f(0)$ is always A_{eq} (a constant).

The constancy of A_{eq} predicted by this model remains valid even for wide cone beams – as clearly illustrated by the experimental data in Table 5.1. This broad general result follows physically from the conservation of energy – the total amount of energy escaping the collimator, impinging on the phantom, and absorbed in the phantom is directly proportional to the aperture a which acts as an energy gate; thus, the constancy of A_{eq} holds along any phantom axis – central or peripheral. Note also that the infinite integrals of the primary beam and scatter components are both (separately) proportional to aperture a, thus a free-in-air measurement (Dixon and Ballard 2007) of $CTDI_\infty$ (air) is proportional to a and thus can provide the relative variation of aperture a with nT (assuming $a <$ integration length in the case of the pencil chamber).

5.4.3.2 Relation Between A_{eq} and the Total Energy Deposited in the Phantom (Integral Dose)

For an axial or helical scan series, the total energy E deposited (absorbed) in the phantom along a given z-axis is represented by the infinite integral of $D_L(z)$ in Eq. (5.1), resulting in

$$E = N \int_{-\infty}^{\infty} f(z)dz = NA_{eq}a;\ \text{the same formula as for the total energy deposited by N rotations}$$

about a stationary phantom (as for SCBCT) for which the dose distribution is $Nf(z)$. With good reason – the total energy deposited by N rotations is independent of their spread or distribution along z and depends (for a given kVp and bow-tie filter) only on the product of $N \times$ (mAs per rotation) $\times$ (aperture a), or simply on the product of (total mAs) $\times$ (aperture a); and the total energy E deposited is the same whether the table moves or not (which likewise applies to the DLP) – see Chapter 2 for energy calculations.

5.4.3.3 Calculation of the Relevant Stationary Phantom Peak Dose f(0) Using this Model

To obtain the desired dose $f(0)$, we evaluate $f(z)$ in Eq. (5.6) at $z = 0$, which is a simple task for the primary beam intensity A_0 which is given by $f_p(0) = A_0$ for the case where $a > c$; however, for the scatter component neither d nor a dominate sufficiently, and the convolution gives a scatter component of,

$$f_s(z) = A_0 \eta \int_{-\infty}^{\infty} d^{-1}h\left(\frac{z-z'}{d}\right)\Pi\left(\frac{z'}{a}\right)dz' = A_0 \eta \int_{-a/2}^{a/2} d^{-1}h\left(\frac{z-z'}{d}\right)dz' \qquad (5.9)$$

Since our goal is to determine the central ray dose $f(0)$ in the stationary phantom which is used as the dose-descriptor for SCBCT, setting $z = 0$ in Eq. (5.9) to obtain $f_s(0)$ and adding the primary beam component $f_p(0) = A_0$ to the scatter component, $f(0) = f_p(0) + f_s(0)$, the total central ray dose is given by,

$$f(0) = f_p(0)\left[1 + \eta \frac{1}{d} \int_{-a/2}^{a/2} h\left(\frac{z}{d}\right) dz\right] \qquad (5.10)$$

If $a \gg d$, the above integral is essentially infinite, and $f(0) = f_p(0)(1 + \eta) = A_{eq}$, corresponding to scatter equilibrium being attained at $z = 0$ (of academic interest only for SCBCT since $d = 117$ mm).

5.4.3.4 The Scatter LSFs Exhibit Surprising Simplicity – the Monte Carlo Model

Further results from this model require a more detailed knowledge of the scatter $LSF(z) = \eta lsf(z)$ on the phantom axis. To that end, Boone (2009) has performed Monte Carlo (MC) dose simulations in a variety of cylindrical phantoms (with and without bow-tie filters); and more importantly has provided the data in its most useful and concise form, namely as a scatter LSF which can be used as a "kernel" in the integral expressions derived above to calculate $f(0)$ [as well as $f(z)$] for any beam width a, *without requiring any additional MC simulations*. Moreover, these scatter LSFs exhibit a surprising simplicity, asymptotically approaching a pure, single- exponential of the form $\exp(-\mu_r z)$ a few cm beyond $z = 0$, thus allowing the theoretical results to be expressed as simple analytical functions.

Figure 5.2 shows the scatter $LSF(z)$ obtained by Boone (2009) for the central axis of a 32 cm diameter PMMA phantom of infinite length at 120 kVp with bow-tie filter, this function being readily fit by a double-exponential decay as shown. Only half of the even function $LSF(z) = LSF(-z)$ has been shown.

Renormalizing the scatter LSF function fit parameters shown in Figure 5.2 to conform to our notation (using scaled, unit-area lsf functions), where $LSF = \eta \times lsf(z)$, it becomes,

$$lsf(z) = (1 - \varepsilon)\frac{1}{d}\exp(-2|z|/d) + \varepsilon\frac{1}{\delta d}\exp(-2|z|/\delta d) \qquad (5.11)$$

where $d = 117$ mm, $\delta d = 6.74$ mm $= .0576d$, and where $(1 - \varepsilon) = 0.985$ and $\varepsilon = 0.015$ are the respective *areas* of the asymptotic first term, and the transient second term; and where the S/P ratio (Boone 2009) is $\eta = 13$ for the central axis of the body phantom. The transient second term in Eq. (5.11) becomes negligible for $z > 10$ mm, after which the lsf reaches its single-exponential asymptotic form, which when written as $\exp(-\mu_r z)$, corresponds to a value of $\mu_r = 0.17$ cm^{-1}.

The transient second term in Eq. (5.11) produces little effect for cone beams (or beams having widths $a > 10$ mm), and the lsf can be approximated quite well in this case by the first term of Eq. (5.11) as a single exponential (re-normalized to unit area), namely as,

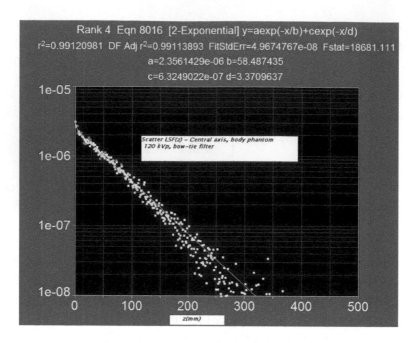

FIGURE 5.2 Scatter LSF, central axis 32 cm diameter PMMA body phantom with double-exponential fit function (arbitrary units). Data from a Monte Carlo simulation (Boone 2009) at 120 kVp with a body bow-tie filter.

(From Dixon and Boone, *Medical Physics*, 2010.)

$$lsf(z) = d^{-1}h(z/d) \cong d^{-1}\exp(-2|z|/d) \tag{5.12}$$

in which $d = 117$ mm.

5.4.3.5 Derivation of the Equation for the Peak Dose f(0) Using the Scatter LSF
Substitution of the simplified Eq. (5.12) into Eq. (5.10) and integrating yields,

$$f(0) \cong f_p(0)[1+\eta(1-e^{-a/d})] \tag{5.13}$$

Using the more accurate double-exponential fit to the *lsf* given *by* Eq. (5.11), which is more appropriate (more accurate) for narrow fan beams, gives,

$$f(0) = f_p(0)\{1+\eta[(1-\varepsilon)(1-e^{-a/d})+\varepsilon(1-e^{-a/\delta d})]\} \tag{5.14}$$

It can be clearly seen from Eqs (5.13 and 5.14) that the growth of the central peak dose $f(0)$ with beam width a is due entirely to an increase in scatter, with the primary beam contribution $f_p(0)$ remaining constant.

Both Eqs (5.13 and 5.14) approach the same limiting dose value when a becomes large compared to d, viz., $f(0) \to f_p(0)[1+\eta] = A_{eq}$, for all practical purposes (within 2%) when $a \geq a_{eq} = 4d = 470$ mm [from $e^{-a/d} = e^{-4} = 0.018$ in Eqs (5.13 or 5.14)]. Although this expression for $A_{eq} = f_p(0)[1+\eta]$ was previously obtained in Eq. (5.7) using the convolution integral theorem; *this model provides a clear physical insight and a functional form (an exponential growth function) describing the approach of f(0) toward its (unattainable) equilibrium value A_{eq}, clearly illustrating that this occurs only in the limit where the cone beam width a itself becomes large with respect to the width d = 117 mm of the scatter LSF, namely, for $a \geq 4d = 470$ mm.* For a primary beam width $a = 4d$, a primary photon scattered at the extreme edges of the beam $z = \pm a/2$ has only a negligible chance of getting back to $z = 0$ to contribute to the center dose $f(0)$ [a survival probability of $\exp(-\mu_r a/2) = e^{-4}$]. The same *LSF* and thus the same argument applies to conventional CT; a primary photon scattered from the extremes of the scan length at $z = \pm L/2$ where $L = 4d$ has exactly the same chance (e^{-4}) of making it back to $z = 0$ to contribute to $D_L(0)$; therefore the equilibrium lengths are identical for the two modalities, i.e., $L_{eq} = a_{eq} = 4d = 470$ mm on the central axis.

Although A_{eq} (like CTDI) does not represent a meaningful or relevant dose value for SCBCT, it has utility as a convenient normalization constant for the *common* approach to equilibrium function $H(\lambda)$ which becomes $H(L)$ for conventional CT and $H(a)$ in stationary phantom CT.

5.4.3.6 The Approach-to-Equilibrium Function H(a)

The relative "approach-to-equilibrium" function $H(a) = f(0)/A_{eq}$ for the stationary phantom using the more accurate Eq. (5.14) is given by,

$$H(a) = \frac{1}{1+\eta} + \frac{\eta}{1+\eta}\left[(1-\varepsilon)(1-e^{-a/d}) + \varepsilon(1-e^{-a/\delta d})\right] \quad (5.15)$$

The first term $1/(1+\eta)$ represents the relative primary beam contribution and the second term the relative scatter contribution; however, since $\delta d = 6.74$ mm, the transient scatter term grows very quickly to its small limiting value $\varepsilon = 0.015$ for $a > 25$ mm, representing only 1.5% of the total scatter at equilibrium.

A simplified form of $H(a)$ applicable to the wider beams of SCBCT is obtained using the single-exponential scatter *lsf* from Eq. (5.12) which produces $f(0)$ in Eq. (5.13), and which in turn gives,

$$H(a) \cong \frac{1}{1+\eta} + \frac{\eta}{1+\eta}(1-e^{-a/d}) \quad (5.16)$$

The second term in both equations represents the scatter contribution which has the form of an exponential growth curve – increasing with beam width a until reaching its asymptotic (equilibrium) limit $H(a) \to 1$ for $a \geq 4d \approx 470$ mm, the fractional scatter contribution at equilibrium being $\eta/(1+\eta)$ and the relative primary contribution $1/(1+\eta)$.

5.4.3.7 The Commonality of the Approach to Equilibrium Function H(a) for Both Stationary Phantom Scanning (e.g., SCBCT) and Conventional Helical or Axial CT Scanning

Note that the same scatter *LSF* function applies both to stationary beam CT and conventional axial or helical CT [being used in both cases to create an axial dose profile $f(z)$ in Eq. (5.6)], however, this profile must additionally be integrated over $(-L/2, L/2)$ for conventional CT to obtain the accumulated dose $D_L(0)$ at $z=0$ [Eq. (5.2)] or $CTDI_L$ [Eq. (5.3)]. The integral theorem [Eq. (5.8)] shows that the equilibrium dose constant $A_{eq} = (b/a)D_{eq} = f_p(0)(1+\eta)$ has the same value for both SCBCT and conventional CT, being independent of aperture a (and scan length L).

For conventional CT, the equilibrium dose D_{eq} also depends on the table increment b (or pitch $p = b/nT$) and aperture a, i.e., $D_{eq} = (a/b)A_{eq}$ is proportional to (a/b) as noted previously. The cone beam dose $f(0)$ necessarily approaches the same equilibrium dose value $f_{eq}(0) = A_{eq}$ with increasing aperture setting a (although it will never get there for practical cone beam widths), as previously asserted and as observed in Figure 5.1.

Since $f(0)_a = D_L(0)$ for $L=a$, then $H(L)A_{eq} = H(a)A_{eq}$, thus $H(L) = H(a)$ for $L=a$, hence a *common function H(λ) applies to both modalities* with $\lambda = L$ or $\lambda = a$.

The function $H(L)$ is essentially independent of aperture (Dixon et al. 2005; Boone 2007), since both D_{eq} and $D_L(0)$ are proportional to $\hat{a}$ [rigorously in the case of D_{eq} and approximately for $D_L(0)$] at least over an aperture range corresponding to $(2.5\ \text{mm} \leq nT \leq 40\ \text{mm})$ (Dixon et al. 2005). Boone (2007) has shown that $H(100\ \text{mm}) = [CTDI_{100}/CTDI_\infty] = [D_{100}(0)/D_{eq}]$ varies by less than 1% over this range of apertures.

5.4.3.8 Comparison of the Theoretical Equation for H(a) with Experiment

Does our *mathematical model of the dose profile $f(z)$*, correctly predict the measured (Mori et al. 2005) variation of $f(0)$ with aperture a given in Table 5.1? Or equivalently, does the derived analytic function $H(a) = f(0)/A_{eq}$ given by Eq. (5.15) [or approximated by Eq. (5.16)] agree with the observed ratio computed from the experimental data (Mori et al. 2005) in Table 5.1? *Figure 5.3 answers this affirmatively* – illustrating the excellent agreement between the theoretical predictions of Eq. (5.15) and Eq. (5.16) (the curves) and the experimental values (Mori et al. 2005) of the relative, stationary phantom peak doses $f(0)_a$ shown by the solid data points (●). *It should be emphasized that this is not empirical curve fitting*, but rather *comparing a physical theory having no adjustable parameters as embodied by Eq. (5.15)* to the observed experimental data, thereby giving added confidence in its general applicability.

The single exponential approximation of the *lsf* in Eq. (5.12) leading to the simple growth curve $H(a)$ of Eq. (5.16) *is indistinguishable* from the double-exponential form of the *lsf* which produces $H(a)$ in Eq. (5.15), also plotted in Figure 5.3 but is obscured by the other curve.

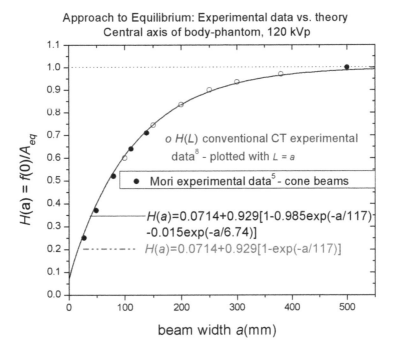

FIGURE 5.3 Approach to scatter equilibrium – theory vs. experiment. Relative approach to equilibrium function $H(a) = f(0)/f_{eq}(0) = f(0)/A_{eq}$ on the central axis of a stationary phantom (e.g., SCBCT). The solid circles (•) are the experimental data of Mori presented in Table 5.1. The solid and dashed lines representing $H(a)$ *are not empirical fits* but rather theoretical predictions obtained from the mathematical model; the solid black line representing the full double- exponential scatter *lsf* of Eq. (5.17), and the essentially congruent dot-dash line using the simpler single – exponential approximation of the scatter LSF of Eq. (5.18). Also plotted is the experimental approach to equilibrium data for conventional CT scans measured (Dixon and Ballard 2007) on GE 16 channel and 64 channel scanners for scan lengths L from 100–400 mm, and plotted using the correspondence $L = a$; thus validating the same functional form for both modalities, $H(L) = H(a)$ with the correspondence $a = L$.

(From Dixon and Boone, *Medical Physics*, 2010.)

The premise that the same function $H(a)$ can also predict the variation of the accumulated dose $D_L(0)$ at $z = 0$ with scan length L for conventional CT simply by making the substitution $L = a$ in Eq. (5.16) is also confirmed by plotting experimental values (Dixon and Ballard 2007) of $H(L)$ measured using helical scans on 16 and 64 channel scanners as described in Chapter 3 (open circles in Figure 5.3) – the data of which also falls on the theoretical curves, thereby confirming this premise that $H(a) = H(L)$, at least over the measurement range used (100 mm $\leq L \leq$ 400 mm). It should be noted, however, that $H(L)$ refers to the variation of dose in a phantom *at least* 500 mm long, and should not be used without correction to extrapolate $CTDI_{100}$ measured in a standard 140 mm long body phantom to predict $CTDI_\infty$, or $D_{eq} = p^{-1} CTDI_\infty$. [e.g., an increase in *body phantom length* from 150 mm to 400 mm was observed (Dixon and Ballard 2007) to produce an increase in the measured value of $CTDI_{100}$ by 7.3% on the central axis and by 1.3% on the peripheral axis at 120 kVp].

5.4.4 Extension to Peripheral Axes

Due to the potential practical utility of the derived analytical functions, it behooves us to make a similar (but abbreviated) analysis for the peripheral axis of the same body phantom (a more complex problem).

That A_{eq} as defined in Eq. (5.5) is likewise a constant (independent of aperture) a on the peripheral axis is confirmed using the experimental data of Mori et al. (2005) as shown in Table 5.2.

The normalized (to unit area) LSFs for the central and peripheral axes are both shown in Figure 5.4 for comparison of the two axes. The *double-exponential fit proved reasonably successful for the peripheral axis*, resulting in the LSF fit parameters of $\varepsilon = 0.304$, $\delta d = 14$ mm, and $d = 88$ mm which, together with $\eta = 1.5$, give a scatter LSF as likewise described by Eq. (5.11) (Figure 5.4).

5.4.4.1 Derivation of the Expression for f(z) and f(0) on the Peripheral Axis using the LSF

As discussed previously (Dixon et al. 2005), the convolution model for $f(z)$ [Eq. (5.6)] strictly holds only for a fixed gantry angle θ on a peripheral axis, since the parameters η, a, c, the *lsf*, and $A_0 = f_p(0)$ are all functions of beam angle θ, thus Eq. (5.6) must be written as $f(z,\theta)$ and then averaged (integrated) over 2π in order to obtain the axial dose profile $f(z)$ on the peripheral axis, thereby losing the convolution format for $f(z)$ and possibly the applicability of the LSF (Boone 2009). The convolution format of Eq. (5.1) for $D_L(z)$ was recovered in the case of conventional helical scanning using this angular average, but this is not particularly relevant here since we want $f(z)$. However, as shown in Appendix A, if $\theta = 0$ denotes the gantry angle for which the beam is directly incident on the peripheral axis in question, then most of the dose on that axis is delivered for a small enough angular range $\pm \Delta\theta$ about $\theta = 0$, such that $a(\theta)$ is slowly varying over $\pm \Delta\theta$ and can be replaced by its average value $\langle a \rangle$, which is the *fwhm* of the axial dose profile $a' = \langle a \rangle$ on the peripheral axis [and which is only about 5% greater than the minimum value of $a(\theta)$ at $\theta = 0$]. Thus, Eq. (5.14) for $f(0)$ and Eq. (5.15) for $H(a)$ also apply, with a replaced by a', and using the corresponding peripheral axis (double-exponential) fit parameters $\varepsilon = 0.304$, $d = 88$ mm, and $\delta d = 14$ mm, and a S/P ratio of $\eta = 1.5$.

TABLE 5.2 Mori Data for the Peripheral Axis of the Body Phantom

Primary Beam Aperture a	Dose Profile Integral $DPI_\infty = \int_{-\infty}^{\infty} f(z)dz$	Equilibrium Dose Constant $A_{eq} = DPI_\infty/a$ $A_{eq} = D_{eq}(a) = \frac{1}{a}\int_{-\infty}^{\infty} f(z)dz$	Measured Central Ray Dose f(0)	Ratio $\frac{A_{eq}}{f(0)}$
138 mm	1520 mGy. mm	11.0 mGy	9.60 mGy	1.14
111 mm	1220	11.0	9.02	1.21
80 mm	900	11.1	8.54	1.28
49 mm	530	10.6	8.06	1.35
28 mm*	290	10.6	7.34	1.49

Source: Dixon and Boone, *Medical Physics* (2010).
*Rounded off from 27.5 mm.

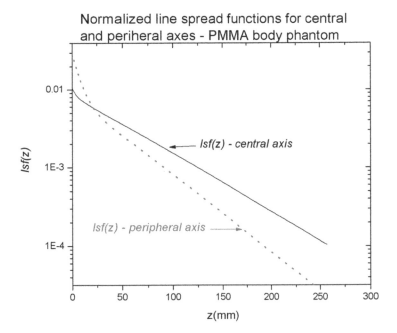

FIGURE 5.4 Scatter *lsf* for the central and peripheral axes in a 32 cm diameter PMMA phantom normalized to unit area. The actual $LSF = \eta \times lsf$, where $\eta = S/P$ ratio with $\eta = 13$ for the central axis and $\eta = 1.5$ for the peripheral axis, thus the absolute peripheral axis *LSF* falls below that for the central axis as expected.

(From Dixon and Boone, *Medical Physics*, 2010.)

$$H(a') = \frac{1}{1+\eta} + \frac{\eta}{1+\eta}\left[(1-\varepsilon)(1-e^{-a'/88\,\text{mm}}) + \varepsilon(1-e^{-a'/14\,\text{mm}})\right] \qquad (5.17)$$

The parameter a', the *fwhm* of the peripheral axis profile $f(z)$ which automatically appears in the peripheral axis equations for $f(0)$ and in $H(a)$, is the physically significant parameter *for comparing modalities* on the peripheral axis; i.e., an axial scan series using beams of width a'/N at a like scan interval $b = a'/N$ with $L = Nb = a'$ will produce a relatively smooth dose distribution (without gaps) on the peripheral axis which is comparable to a cone beam distribution $f(z)$ of width a'. Figure 5.5 shows the relative stationary phantom, peripheral axis peak dose experimental data (Mori et al. 2005) in Table 5.2 plotted vs. $a' = 0.76a$ compared to the theoretical curve $H(a')$ in Eq. (5.17). Likewise, the experimental conventional helical peripheral axis CT data (Dixon and Ballard 2007) $H(L)$ is plotted using the correspondence $L = a'$, with both data sets exhibiting reasonably good agreement between experiment and the theoretical curve $H(a')$ in Eq. (5.17).

Since a' is proportional to the central axis aperture projection a, either could be used to evaluate A_{eq} as noted in Appendix A (it is independent of aperture), and it is more convenient to work with a and $H(a)$ as was done in Table 5.2. By substituting $a = a'/0.76$ into $H(a')$ we obtain,

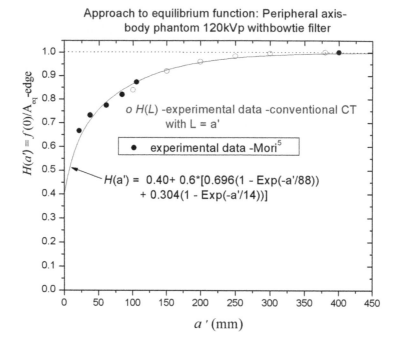

FIGURE 5.5 Approach to equilibrium function $H(a')$ on the peripheral (edge) axis of the body phantom plotted vs. the *fwhm a'* of the cone beam dose profile $f(z)$ on that axis, together with the experimental data of Mori and helical scan data plotted using $L = a'$.

(From Dixon and Boone, *Medical Physics*, 2010.)

$$H(a) = \frac{1}{1+\eta} + \frac{\eta}{1+\eta}\left[(1-\varepsilon)(1-e^{-a/116}) + \varepsilon(1-e^{-a/18.4})\right] \qquad (5.18)$$

Where now $d = 88/0.76 = 116$ mm, $\delta d = 14/0.76 = 18.4$ mm, with $\varepsilon = 0.305$ and $\eta = 1.5$ as before. (It is interesting that $d = 116$ mm which determines asymptotic equilibrium is essentially the same as on the central axis where $d = 117$ mm). Of course, $H(a)$ in Eq. (5.18) is completely equivalent to $H(a')$ in Eq. (5.17) in which $d = 88$ mm and $\delta d = 14$ mm.

For comparison with conventional CT, we must still substitute $L = a'$ into $H(a')$ in order to obtain,

$$H(L) = \frac{1}{1+\eta} + \frac{\eta}{1+\eta}\left[(1-\varepsilon)(1-e^{-L/88}) + \varepsilon(1-e^{-L/14})\right] \qquad (5.19)$$

with $d = 88$ mm, $\delta d = 14$ mm, and $\varepsilon = 0.305$, and $\eta = 1.5$, where $H(L) = D_L(0)/D_{eq}$ and $D_{eq} = (a/b)A_{eq}$.

The reason Eqs (5.17 and 5.18) differ in form but not substance is that the scan length L does not diverge (it is the same for the central and peripheral axes), whereas the beam width a does diverge – we have chosen to work with a rather than a' simply for convenience. Note that dose equilibrium is approached to within 2% on the peripheral axis (our central axis criterion) for $a'_{eq} = L_{eq} = 300$ mm, and $a_{eq} = 400$ mm (compared to $L_{eq} = a_{eq} = 470$ mm on the central axis). A more complete set of $H(L)$ data for various kVp, beam filters, and phantom diameters have been published by Li et al. (2011, 2012a,b, 2013, 2014) including parametric fit parameters.

5.5 GENERATING THE COMPLETE DATA SET FOR CONVENTIONAL AND STATIONARY PHANTOM CT FROM A SINGLE MEASUREMENT OF THE PEAK DOSE $f(0)$ RESULTING FROM A SINGLE AXIAL ROTATION – AN EXAMPLE

You don't really need to scan a long phantom with a helical (or axial) scan series to get $CTDI_L$ or $D_L(0)$ for any scan length.

For added clarity, the peak axial profile dose $f(0)$ is parameterized in the following with the aperture a (primary beam width) as $f(0)_a$ and the experimental data of Mori et al. (2005) for a 32 cm diameter PMMA body phantom are utilized. We will presume for illustrative purposes that only a single measurement of $f(0)_a$ has been made at a single aperture setting a, using a single axial rotation about a small ion chamber (e.g., a Farmer-type chamber) in a stationary phantom. The strategy is then to determine the value of A_{eq} by inverting the relation,

$$f(0)_a = H(a)A_{eq} \qquad (5.20)$$

and applying the analytical functions derived for $H(a)$ in Eqs (5.15 or 5.18). Having obtained A_{eq} we can then predict the desired *stationary phantom* dose $f(0)_a$ for any aperture a from Eq. (5.20) – for the wide cone beams of SCBCT and the narrow fan beams of perfusion studies (or any procedures using a stationary phantom).

But recall we also showed that the same equilibrium dose A_{eq} applies to conventional axial or helical CT using phantom/table motion for a scan interval $b = â$ or a pitch $p = â/nT$. Thus, we can parlay the peak dose $f(0)_a$ value acquired during a single axial rotation about a stationary phantom, as measured using a conventional short ion chamber, into the conventional CT equilibrium dose D_{eq} attained for any aperture setting $â$ (any nT) for any table increment b (or any pitch $p = b/nT$) using $D_{eq} = (â/b)\, A_{eq}$, and thence the dose for any scan length L from $D_L(0) = H(L)\, D_{eq}$ using the same function expressed as $H(L)$ – *the complete data set from a single axial rotation without ever scanning over L.*

The following example using the data of Mori et al. (2005) in Tables 5.1 and 5.2, for the central and peripheral axes of the body phantom, respectively, will be used to illustrate the feasibility of this somewhat ambitious plan. It is assumed that the scanner can be operated in both a conventional CT mode as well as the SCBCT mode. The key to success is an accurate determination of A_{eq} which suggests using a reasonably wide beam – such as $a = 111$ mm. We therefore presume that only a single measurement of the peak axial dose $f(0)_a = f(0)_{111}$ is made on each phantom axis, these being the values shown in bold type in Table 5.3 for $a = 111$ mm. *We will then attempt to predict the other data from* these [and then check our predictions (shown in italics) against the actual measured data]. This is best illustrated by displaying the results in tabular form as in Table 5.3.

TABLE 5.3 Stationary Phantom Doses $f(0)_a$ – as Predicted from a Measurement of Peak Dose for a Single Aperture Setting

	Central Axis				Peripheral Axis			
	Peak Dose $f(0)_a$				Peak Dose $f(0)_a$			
a(mm)	Eq. (5.15) $H(a)$	Predicted $H(a)A_{eq}$	Measured Table 5.1	Error	Eq. (5.18) $H(a)$	Predicted $H(a)A_{eq}$	Measured Table 5.2	Error
138 mm	0.715	4.35 mGy	4.37 mGy	0.5%	0.873	9.38 mGy	9.60 mGy	−2.3%
111 mm	0.641		**3.90**		0.84		**9.02**	
49 mm	0.385	2.34	2.27	3.1%	0.714	7.67	8.06	−5.1%
28* mm	0.272	1.63	1.53	8%	0.630	6.74	7.34	−9.0%
7.8 mm	0.140	0.85			0.490	5.26		
∞	1.000	$A_{eq} = 6.08$	$A_{eq} = 6.17$	1.5%	1.000	$A_{eq} = 10.7$	$A_{eq} = 10.9$	−1.5%

Source: Dixon and Boone, *Medical Physics* (2010).

In this example, $f(0)_{111}$ was used as the measured value (shown in bold letters) from which the values of A_{eq} shown in the last row were predicted using $H(a)$. Predicted doses are given in *italics*, the actual measured doses (Mori et al. 2005) from Tables 5.1 and 5.2 are also listed for comparison, and the % error is shown.

* Rounded off from 27.5 mm.

The value of the aperture-independent constant A_{eq} is *predicted* using the measured peak dose for $a = 111$ mm using $A_{eq} = f(0)_a/H(a) = f(0)_{111}/H(111) = 3.90/0.641 = 6.08$ mGy on the central axis, and $A_{eq} = 9.02/0.84 = 10.74$ mGy on the peripheral axis, which agree (within 1.5%) with the average experimental values of 6.17 and 10.86 from Tables 5.1 and 5.2. Thence the value of the stationary phantom peak dose $f(0)_a = H(a)A_{eq}$ can be predicted for any aperture a, and some representative values are given in Table 5.3 for comparison with the measured values where available.

It might also be prudent to measure $f(0)_a$ for several values of a, since the additional acquisition time is minimal and of little consequence, and likewise to measure a $CTDI_L$ to confirm a good crossover. The doses for a *narrow aperture* of $a = 7.8$ mm ($nT = 5$ mm) such as one might use in a stationary phantom perfusion study are also shown in Table 5.3, to illustrate that $CTDI_{100}$ over-estimates the peak dose $f(0)$ by factors of 4.7 and 2.0 on the central and peripheral axes, respectively – and thus the reported $CTDI_{vol}$ over-estimates the *peripheral axis peak dose* (comparable to the skin dose) by a factor of 2.3 (see Bauhs et al. 2008).

5.5.1 Crossover to Conventional CT dose

We can also extend this single measurement of the SCBCT peak dose $f(0)_{111}$ to conventional CT and generate complete dose tables for that modality. We already have the predicted value of the quantity of greatest interest (and the most difficult to measure) in helical or axial scanning – namely the equilibrium dose constant A_{eq} (with predicted values of 6.08 and 10.7 mGy from Table 5.3), which has the same value for conventional CT (assuming the same bow-tie filter) and from which the conventional CT equilibrium dose D_{eq} for any table increment b (or any pitch $p = b/nT$) and for any fan beam aperture â (or nT) can be obtained using $D_{eq} = (\hat{a}/b) A_{eq} = (\hat{a}/nT)p^{-1}A_{eq}$, and thence the dose for any scan length L can be obtained using $D_L(0) = H(L) D_{eq}$. For example, assuming the scanner is operated in a helical mode with $nT = 64 \times 0.5$ mm $= 32$ mm (corresponding to an aperture â $= 49$ mm), pitch $p = 1.35$, $\tau = 1$ sec, total beam-on time $t_0 = 23$ sec, scan length $L = vt_0 = 100$ mm; then $D_{eq} = (\hat{a}/b)A_{eq} = (/\hat{a} nT)p^{-1}A_{eq} = 1.13A_{eq} = 6.9$ and 12.1 mGy on the central and peripheral axes, respectively; which, when corrected to $L = 100$ mm using the values of $H(L) = 0.60$ from Eq. (5.17) and $H(L) = 0.86$ from Eq. (5.19), results in dose values of $D_L(0) = H(L)D_{eq} = 4.1$ and 10.4 mGy on the two axes.

5.6 SUMMARY AND CONCLUSIONS

- For SCBCT scanning (without table/phantom motion), it was shown that the peak dose $f(0)$ on the central ray ($z = 0$) of the cone beam is the logical (and unique) choice for a SCBCT dose-descriptor consistent with the CTDI-based dose used in conventional CT.

- This "point-dose" $f(0)$ can be directly measured using a single axial rotation about a small ion chamber located in the phantom on the central ray of the primary beam ($z = 0$).

- A common measurement method can be utilized for both the stationary and moving phantom, viz., a conventional, short ionization chamber (such as a 0.6cc Farmer-type chamber) located at the center $z=0$ of the directly irradiated length to measure the dose $Nf(0)$ for SCBCT, or the accumulated dose $D_L(0)$ at $z=0$ in conventional CT using a helical (or axial) scan series as illustrated in Chapter 3 (Dixon and Ballard 2007) to translate the phantom (and ion chamber) over $(-L/2, L/2)$ where $L=\upsilon t_0$ (or $L=Nb$). *This direct measurement method is actually more general than the theoretical equations, requiring neither shift-invariance of the phantom, the x-ray beam, nor the scan interval.*

- The equilibrium dose constant $A_{eq}=(b/a)D_{eq}$ is independent of both pitch $p=b/nT$ and aperture a (thence nT), and was shown to have a common value for both SCBCT (stationary cone beam CT) and conventional helical or axial CT, and its constancy has been demonstrated over a wide aperture range (28 mm $\leq a \leq$ 138 mm); therefore it is sufficient to determine A_{eq} at a single known aperture value a (which needn't be a wide beam).

- Many common features of the SCBCT dose $f(0)$ as a function of cone beam width a and the conventional CT dose $D_L(0)$ as a function of scan length L have been established, including a common equilibrium dose constant A_{eq}, a common scatter-equilibrium length $a_{eq}=L_{eq}$, and a common function $H(\lambda)$ which describes the relative approach to dose equilibrium for both modalities, where $\lambda=a$, or $\lambda=L$, such that $f(0)_a = H(a)A_{eq}$, and $D_L(0) = H(L)D_{eq} = H(L)(b/a)A_{eq}$.

- Using the scatter *LSF* derived from the Monte Carlo simulation of Boone (2009), analytic functions describing the variation of the peak dose of an axial dose profile $f(0)_a$ and $H(a) = f(0)_a / A_{eq}$ as a function of collimator aperture a (primary beam width) were derived [e.g., see Eqs (5.14 and 5.15)] which functions provided a good match to the experimental data (Mori et al. 2005; Dixon and Ballard 2007), and which have importance and utility for predicting the peak dose for the narrow fan beams used in perfusion studies (Bauhs et al. 2008) as well as the relevant cone beam dose $f(0)$ in SCBCT for any beam width (aperture) a.

- The commonality described also suggests the possibility that a single measurement of the peak dose $f(0)_a$ of an axial dose profile resulting from a single rotation about a stationary phantom for a single (arbitrarily chosen) aperture a setting using a small ion chamber is sufficient to predict the "peak" dose $f(0)_a$ for any other aperture – for wide cone beams and narrow fan beams alike, including the scatter-equilibrium dose constant A_{eq}, using the function $H(a)$ to obtain $A_{eq}=f(0)_a/H(a)$. The value of A_{eq} can then be used to predict the equilibrium dose D_{eq} for conventional axial or helical CT scans for any table increment b (or any pitch $p=b/nT$) at any aperture setting a (any nT) using $D_{eq}=(a/b)\,A_{eq}$, and thence to predict the dose for any desired sub-equilibrium scan length L using $D_L(0)=H(L)\,D_{eq}$ using the same analytical function $H(\lambda)$ with $\lambda=L$; obtaining the complete data set without ever scanning over L.

That is, the theory developed allows a "crossover" between stationary phantom and conventional (helical or axial) scanning modalities (assuming shift invariance exists).

In fact, one obtains the complete data set for both modalities, namely $f(0)_a$ for any a, A_{eq}, $D_{eq} = (a/b) A_{eq}$, and $D_L(0)$ for any L – all from a single measurement of the peak dose $f(0)_a$ of an axial dose profile resulting from a single axial rotation about a stationary phantom.

- Inspection of Figure 5.3 illustrates the rather remarkable confluence of an analytical, theoretical function based on a scatter *LSF* obtained from a Monte Carlo simulation, with *measured* axial profile peak doses $f(0)_a$ measured on a Toshiba 256 channel cone beam scanner, and with helical scan doses $D_L(0)$ measured on conventional GE LS scanners, which speaks to the generality of these results.

- This work is based on Monte Carlo data (Boone 2009) obtained at 120 kVp in a 32 cm diameter PMMA phantom using a typical bow-tie filter, and is intended as proof of concept exposition which successfully matched the experimental data (Mori et al. 2005; moreover, the theory and equations are quite general and can now be easily extended (Boone 2009) to any phantom diameter (for a variety of phantom materials) at a variety of kVp settings (with or without) bow-tie filters. These systematics have been developed (Li et al. 2011; 2012; 2013) in which $H(L)$ values have been derived for a wide variety of conditions (kVp, beam filters, phantom diameters, etc.). One particular noteworthy accomplishment (Li et al. 2012) was developing the equations [Eqs (5.21 and 5.22)] for the dose at any point inside or outside the scan length (via a clever derivation). That is, using $H(L) = D_L(0)/D_{eq}$.

It was shown by Li et al. (2012) that the dose at any point inside the scan length ($-L/2$, $L/2$) could be written as,

$$D_L(z) = \frac{1}{2} D_{eq} \left[H(L + 2z) + H(L - 2z) \right], |z| < L/2 \qquad (5.21)$$

And the dose at a distance $z_0 = z - L/2$ beyond the scan length as,

$$D_L(z) = \frac{1}{2} D_{eq} \left[H(2L + 2z_0) - H(2z_0) \right] \quad |z| > L/2 \qquad (5.22)$$

which can also be expressed in terms of $D_L(0)$ instead of D_{eq} using $H(L) = D_L(0)/D_{eq}$.

ACKNOWLEDGMENT

The author is grateful to Dr. Shinichiro Mori for supplying the numerical beam profile data for the various cone beam widths for the prototype 256 channel scanner as previously published in *Medical Physics* (Mori et al. 2005)

APPENDIX A: DERIVATION OF THE LSF FORMULATION FOR THE PERIPHERAL AXIS

By analogy with Eq. (5.6),

$$f(z, \theta) = A_0(\theta) \left[\frac{1}{c(\theta)} g\left(\frac{-z}{c(\theta)} \right) + \eta(\theta) \mathrm{lsf}(z, \theta) \right] \otimes \Pi\left(\frac{z}{a(\theta)} \right) \qquad (A.1)$$

where $A_0(\theta) = f_p(0,\theta)$ is the primary beam intensity (dose rate) on the peripheral axis, and is thus the appropriate weighting function to average various parameters over θ as related to the peripheral axis (Dixon et al. 2005). For example, the average of $A_0(\theta)a(\theta)$ can be written as $A_0\langle a\rangle$ where $A_0 = f_p(0)$ is the *total* primary beam dose on the peripheral axis [the integral of $A_0(\theta) = f_p(0,\theta)$ over all angles (Dixon et al. 2005)], and $\langle a\rangle$ is the dose-weighted average of the projected aperture $a(\theta)$ on the peripheral axis, which is equal to the width (*fwhm*) of the primary beam profile $f_p(z)$ on that axis. The presence of $a(\theta)$ in the rect function $\Pi[z/a(\theta)]$ in Eq. (A.1) breaks the *shift-invariant* symmetry required by the convolution once the integration over θ is performed. Expressing the convolution in Eq. (A.1) in its integral form, illustrates the difficulty, namely,

$$f(z,\theta) = A_0(\theta)\left[1 + \int_{-a(\theta)/2}^{a(\theta)/2} \mathrm{LSF}(z-z',\theta)dz'\right] \tag{A.2}$$

Fortunately, the problem is solvable in closed form on the peripheral axis, since most of the dose on a given peripheral axis is contributed while the beam is directly incident on that axis ($\theta = 0$), due to the fact that phantom, bow-tie filter, and secondarily inverse square attenuation (Dixon et al. 2005) serve to rapidly "pinch off" the primary beam intensity $A_0(\theta)$ at angles beyond about $\pm 50°$, over which angular range $a(\theta)$ is slowly varying. This likewise applies to the peripheral axis of the head phantom (Dixon et al. 2005), where the rolloff of $A_0(\theta)$ is slower, but which is compensated by a smaller variation of $a(\theta)$. In fact, it was previously shown in Chapter 4 (Dixon et al. 2005) that $\langle a\rangle = 1.05\,a(0)$ for the peripheral axes in both body and head phantoms and only 5% above its minimum value $a(0)$ at $\theta = 0$. This limited variation of $a(\theta)$ allows one to replace it in Eqs (A.1 and A.2) by its average value $a' = \langle a\rangle$ with negligible error, thereby preserving the convolution format (a' is equal to the *fwhm* of the axial dose profile on the peripheral axis). Indeed, this approximation is good on a z-axis located at any radius in either phantom.

Replacing $a(\theta)$ by $a' = \langle a\rangle$ in Eq. (A.2) and setting $z=0$ gives,

$$f(0,\theta) = A_0(\theta)\left[1 + \int_{-a'/2}^{a'/2} \mathrm{LSF}(z',\theta)dz'\right] \tag{A.3}$$

Which, when averaged (integrated) over θ, gives,

$$f(0) = A_0 + \int_{-a'/2}^{a'/2} dz' \frac{1}{2\pi}\int_{-\pi}^{\pi} A_0(\theta)\,\mathrm{LSF}(z',\theta)d\theta = A_0 + A_0\int_{-a'/2}^{a'/2}\langle \mathrm{LSF}(z')\rangle dz' \tag{A.4}$$

where $\langle \mathrm{LSF}(z')\rangle$ denotes the dose-weighted, angular average of $\mathrm{LSF}(z,\theta) = \eta(\theta)\mathrm{lsf}(z,\theta)$ over a complete rotation.

Eq. (A.4) is seen to have the same form as Eq. (5.11) for the central axis [recognizing the scaled form of the *LSF* previously used $LSF(z) = \eta d^{-1}h(z/d)$, thus Eq. (5.16) for $f(0)$ and Eq. (5.17) for $H(a)$ also apply, if one replaces a with the *fwhm* of the axial dose profile on the peripheral axis $a' = \langle a \rangle \approx 1.05a(0) \approx 0.76a$, where $a(0) = a(F - 15 \text{ cm})/F$, and where F is the focal to AOR distance, and uses the values of the (double-exponential) fit parameters (ε, d, δd) appropriate to the peripheral axis and $\eta = 1.5$.

GLOSSARY

MDCT: multi-detector CT

SCBCT: stationary phantom cone beam CT

Fan beam: a nominal beam width of ≤ 40 mm along z (≥ 40 mm is typically called a cone beam)

Shift-invariance: translational invariance (independent of location along the z-axis)

AOR: gantry axis of rotation located at isocenter; $F =$ source to isocenter distance

t_0: total beam-on time for an axial or helical scan series (tube loading time)

τ: time for single 360° gantry rotation (typically $\tau = 1$ sec or less)

$N = (t_0/\tau)$: total number of gantry rotations in an axial or helical scan series (N may not be an integer for helical scanning)

v: table velocity for helical scans

b: generalized table advance per rotation (mm/rot), or *table index*

$b = v\tau$: for helical scans; $b =$ scan interval for axial scans (denoted elsewhere as Δd or "I")

$L = vt_0$: definition of total helical scan length (the total reconstructed length is $< L$)

$L = Nb$: generalized definition of total scan length (axial or helical)

$\Pi(z / L)$: rectangular function of unit height and width L spanning interval $(-L/2, L/2)$

nT: total slice width acquired in a single rotation (often denoted by "N×T"). Also equal to the total active detector length projected at isocenter for MDCT (e.g., $nT = 16 \times 1.25$ mm $= 20$ mm)

a: geometric projection of the z-collimator aperture onto the AOR (by a "point" focal spot). For MDCT $a > nT$ (called "over-beaming") to keep penumbra beyond active detector length nT

$p = b/nT$: generalized "pitch"

Accumulated dose: dose accrued at a given z (e.g., z = 0) due to a complete series of N axial or helical rotations

$f(z)$: single rotation (axial) dose profile acquired with the phantom held stationary

D_{eq}: limiting value of accumulated dose approached in conventional CT for scan lengths $L \geq L_{eq}$

L_{eq}: scan length required for dose to approach to within $< 2\%$ of D_{eq} at $z = 0$ (denoted symbolically as $L \rightarrow \infty$)

A_{eq}: the equilibrium dose constant, equal to D_{eq} for a table increment $b = a$ (and independent of aperture a and nT)

R: radius of cylindrical phantom

η: scatter to primary ratio S/P

The following development concentrates on the *actual in-phantom dose* for both conventional CT involving phantom translation, and stationary phantom CT such as SCBCT, directed toward creating a consistent approach to CT dose assessment which provides continuity of dose and physical interpretation between these two modalities. To that end, it is necessary to review the dose-descriptors used in conventional CT (and the CTDI paradigm).

REFERENCES

AAPM 2010, Report of AAPM Task Group 111, Comprehensive methodology for the evaluation of radiation dose in x-ray computed tomography, American Association of Physicists in Medicine, College Park, MD, February (2010), http://www.aapm.org/pubs/reports/RPT_111.pdf.

Anderson J., Chason D., Arbique G., and Lane T., New approaches to practical CT dosimetry. Abstract SU-FF-I-14: *Med Phys* 32, 1907, (2005).

Bauhs J.A., Vrieze T.J., Primak A.N., Bruesewitz M.R., and McCollough C.J., CT dosimetry: Comparison of measurement techniques and devices. *Radiographics* 28, 245–253, (2008).

Boone J.M., Dose spread functions in computed tomography: A Monte Carlo study. *Med Phys*, In press, (2009).

Boone J.M., The trouble with $CTDI_{100}$. *Med Phys* 34, 1364, (2007).

Bracewell R.N., *The Fourier Transform and Its Applications*, 3rd ed., McGraw Hill, Boston, (2000).

Dixon R.L., A new look at CT dose measurement: Beyond CTDI. *Med Phys* 30, 1272–1280, (2003).

Dixon R.L., and Ballard A.C., Experimental validation of a versatile system of CT dosimetry using a conventional ion chamber: Beyond $CTDI_{100}$. *Med Phys* 34(8), 3399–3413, (2007).

Dixon R.L., and Boone J., Cone beam CT dosimetry: A unified and self-consistent approach including all scan modalities – With or without phantom motion. *Med Phys* 37, 2703–2718, (2010).

Dixon R.L., Munley M.T., and Bayram E., An improved analytical model for CT dose simulation with a new look at the theory of CT dose. *Med Phys* 32, 3712–3728, (2005).

Fahrig R., Dixon R.L., Payne T.L., Morin R.L., Ganguly A., and Strobel N., Dose and image quality for a cone-beam C-arm CT system. *Med Phys* 33, 4541–4550, (2006).

Leitz W., Axelson B., and Szendro G., Computed tomography dose assessment – A practical approach. *Radiat Prot Dosim* 57, 377–380, (1995).

Li X., Zhang D., and Liu B., A practical approach to estimate the weighted CT dose index over an infinite integration length. *Med Phys Biol* 56, 5789–5803, (2011).

Li X., Zhang D., and Liu B., Estimation of the weighted $CTDI_{\infty}$ for multislice CT examinations. *Med Phys* 39, 901–905, (2012a).

Li X., Zhang D., and Liu B., Equations for CT dose calculations on axial lines based on the principle of symmetry. *Med Phys* 39, 5437–5352, (2012b).

Li X., Zhang D., and Liu B., Monte Carlo assessment of CT dose equilibration in PMMA and water cylinders with diameters from 6 to 55 cm. *Med Phys* 34, 3399–3413, (2013).

Li X., Zhang D., and Liu B., Longitudinal dose distribution and energy absorption in PMMA and water cylinders undergoing CT scans. *Med Phys* 41(10), (2014).

Mori S., Endo M., Nishizawa K., Tsunoo T., Aoyama T., Fujiwara H., and Murase K., Enlarged longitudinal dose profiles in cone-beam CT and the need for modified dosimetry. *Med Phys* 32, 1061–1069, (2005).

Nakonechny K.D., Fallone B.G., and Rathee S., Novel methods of measuring single scan dose profiles and cumulative dose in CT. *Med Phys* 32, 98–109, (2005).

Shope T., Gagne R., and Johnson G., A method for describing the doses delivered by transmission x-ray computed tomography. *Med Phys* 8, 488–495, (1981).

CHAPTER 6

Analytical Equations for CT Dose Profiles Derived Using a Scatter Kernel of Monte Carlo Parentage Having Broad Applicability to CT Dosimetry Problems

6.1 INTRODUCTION

6.1.1 Summary of Pertinent Results from the Previous Chapter

This chapter is an extension of Chapter 5 in which a Monte Carlo generated scatter *LSF* [also referred to as the *dose spread function (DSF)* by Boone (2009)] was used as a kernel in a convolution-based model to generate an analytic equation describing the peak dose $f(0)$ of an axial dose profile $f(z)$ as a function of primary beam width a, including both wide cone beams and narrow fan beams (where a is the z-collimator aperture projected at isocenter). The interest in the peak dose $f(0)$ is due to the fact that it was shown in Chapter 5 to be the logical (and unique) choice for a stationary phantom cone beam (SCBCT) dose-descriptor consistent with the CTDI-based dose used in conventional CT. AAPM Report 111 (AAPM 2010) likewise recommends using $f(0)$ for stationary phantom CT.

6.1.2 Deriving an Analytical Function Describing the Complete Dose Profile

The primary objective of the present chapter is to derive an analytical function describing the complete dose profile $f(z)$ resulting from a *single axial rotation*, by further exploiting the same scatter $LSF(z) = \eta\, lsf(z)$ obtained from the MC simulation (Boone 2009) expressed in analytical form (Dixon and Boone 2010) using the same convolution model as in Chapter 5, expressed in Eq. (6.1) as the sum of primary and scatter components,

$$f(z) = f_p(z) + f_s(z) = \left[\frac{1}{c} g \left(\frac{-z}{c} \right) + \eta \mathrm{lsf}(z) \right] \otimes A_0 \Pi \left(\frac{z}{a} \right) \qquad (6.1)$$

Where $(A_0/c)g(-z/c)$ represents the focal spot emission intensity as slit projected (by the z-collimators) onto the axis of rotation as shown in Chapter 4 (Dixon et al. 2005). Note that Eq. (6.1) ignores any *scatter asymmetry* which might result from perturbation of the primary beam profile due to the heel effect, but the primary beam function includes both the heel effect and asymmetric penumbra due to anode-tilt as developed in Chapter 4. Achieving this primary objective alone; namely, deriving an analytic expression for $f(z)$ as a function of aperture a from basic physical principles with all parameters *fixed* by the physics (without the use of any empirical functions or arbitrary, adjustable "fit" parameters), and the function of which further exhibits reasonable congruence with the experimental profile data, is an ambitious undertaking. It is emphasized that fitting CT beam profiles using arbitrary functions containing adjustable "fit" parameters is merely a mathematical exercise, lacking the physical basis for establishing its limits of applicability or its generality, in contrast to this approach in which a poor match between theory and experiment allows little recourse – the parameters are not adjustable. However, a successful match provides a powerful tool for dose simulations and for facilitating further physical insight. With a complete analytical function $f(z)$, one can generate (simulate) the CT dose distribution in a cylindrical phantom under practically any conditions.

For stationary phantom CT, cone beam (SCBCT) and fan beam alike, the axial dose profile $f(z)$ itself describes the complete dose distribution and its central value at $z = 0$. $f(0)$ is the signature dose value of interest (Dixon and Boone 2010) for SCBCT consistent with the CTDI for conventional axial or helical scanning as well as for perfusion studies (Dixon and Boone 2010; AAPM 2010; Bauhs et al. 2008; Fahrig et al. 2006); hence $f(z)$ by itself is the dose distribution for SCBCT, requiring no further operations on (or by) $f(z)$ such as integration or convolution (and CTDI does not apply).

However, conventional helical or axial CT scanning may require further operations on the fan beam profile $f(z)$ such as convolution to obtain the accumulated dose distribution $D_L(z)$ in Eq. (6.3) or integration over the scan length L in order to obtain the central dose at $z = 0$, $D_L(0) = [CTDI_L/\mathrm{pitch}]$ in Eq. (6.4) and related parameters. In fact, an analytical expression for $CTDI_L$ is derived herein, directly illustrating its physical underpinnings.

6.1.3 The Utility of the Analytical Dose Profile Function $f(z)$

Calculation of the actual accumulated dose distribution in conventional CT using $f(z)$ *does not require* integration over $f(z)$, rather the dose distribution for an axial scan series $\tilde{D}_N(z)$ is obtained by summing the dose resulting from $N = 2J + 1$ axial scans spaced at equal intervals b along z (where b is the table advance between rotations); the sum over the progressively displaced profiles being given by

$$\tilde{D}_N(z) = \sum_{k=-J}^{J} f(z - kb) = f(z) \otimes \sum_{-J}^{J} \delta(z - kb) \qquad (6.2)$$

where $\widetilde{D}_N(z)$ is, in general, quasi-periodic (oscillatory) of fundamental period b, and is readily generated if the *complete function f(z)* is known [including its scatter tails down to about 1% of the peak dose $f(0)$]. This has not previously been possible with the degree of generality provided by the analytic expression for $f(z)$ derived herein.

The CTDI-paradigm is based on the philosophy that averaging out any such periodic dose oscillations such as produced by Eq. (6.2) is desirable; and it has been shown in Chapter 2 that the dose distribution, smoothed by averaging over one period of rotation, is given by the convolution (Dixon 2003),

$$D_L(z) = \frac{1}{b} f(z) \otimes \Pi(z/L) = \frac{1}{b} \int_{-L/2}^{L/2} f(z-z')dz' \qquad (6.3)$$

The value of this smoothed dose accumulated at the center of the scan length $(-L/2, L/2)$ is obtained by setting $z = 0$ in Eq. (6.4), namely,

$$D_L(0) = \frac{1}{b} \int_{-L/2}^{L/2} f(z')dz' = p^{-1}\mathrm{CTDI}_L \qquad (6.4)$$

where $CTDI_L$ itself represents the value of the smooth dose at $z = 0$ for a particular value of table advance $b = nT$ (or a pitch $p = b/nT = 1$).

For axial scans, smoothing of $\widetilde{D}_N(z)$ in Eq. (6.2) is accomplished by taking its "running mean" (Dixon 2003) – a longitudinal average over one period of oscillation $(-b/2 \leq z \leq b/2)$ at each z, and obtainable (Bracewell 2000) by convolving Eq. (6.2) with $b^{-1}\Pi(z/b)$.

For helical scanning, the physical interpretation of the smoothed doses $D_L(z)$ and $D_L(0)$ in Eqs (6.3 and 6.4) and for $CTDI_L$ itself *is different* as shown in Chapter 2 and described below (Dixon 2003).

1. *On the central phantom axis,* Eq. (6.3) represents the absolute, point-dose distribution which is naturally smooth (non-oscillatory) with *no averaging being required,* and Eq. (6.4) gives the absolute, point-dose at $z = 0$.

2. *On the peripheral phantom axes,* Eqs (6.3 and 6.4) represent an angular average of the helical dose distribution over 2π at a fixed value of z (an average over all peripheral axes at fixed z).

Advantages of Dose Profile Dosimetry over CTDI

1. Knowledge of the dose profile $f(z)$ allows one to compute the accumulated dose at any point along z using Eq. (6.2) or Eq. (6.3), whereas the CTDI formalism (including $CTDI_{vol}$) only provides the dose at the center $(z=0)$ of the scan interval $(-L/2, L/2)$, viz., $D_L(0) = p^{-1} CTDI_L$ using Eq. (6.3).

2. Freedom from the constraints of *shift-invariance* required for the predictive CTDI formalism to apply (Dixon and Boone 2010; Dixon et al. 2005) and as illustrated in Chapter 2. The CTDI method uses the integral of a single axial dose profile (typically

acquired using a pencil chamber) to predict the accumulated dose at the center of the scan length ($z = 0$) resulting from N *identical* dose profiles spaced at *equal intervals* along z (*a constant pitch*); whereas our analytical profile functions $f(z)$ can be individualized, allowing dose calculations for variable apertures, variable mA, irregular spacing along z (or variable pitch) to mention a few; and also permits dose calculations at any desired point along z (not just $z = 0$).

A further objective is extending the analytical model to cover SCBCT by demonstrating the efficacy of the theory in matching the experimental beam profile data $f(z)$ (the complete dose distribution for SCBCT) for wider cone beams ($a > 40$ mm).

6.2 MATERIALS AND METHODS

All simulations and numerical parameters refer to a 32 cm diameter PMMA "body" phantom (at least 570 mm in length), scanned at 120 kVp with bow-tie filter; however, the general methodology and theory is broadly applicable to any cylindrical, dosimetry phantom for any kVp.

The experimental beam profile data (Mori et al. 2005) from a 256 channel prototype cone beam CT scanner were used to corroborate the theory for beam widths ranging from 28 mm up to 138 mm. Peripheral axis dose distributions were also measured (Dixon et al. 2005; Dixon and Ballard 2007) using Kodak EDR2 film, digitized using a scanner as previously described in Chapter 4 (Dixon et al. 2005) with additional data obtained using 15 cm long Landauer OSL strips (Dixon and Ballard 2007), for comparison with the theoretical simulated dose (these data were acquired with sub-mm resolution).

6.3 THEORY

6.3.1 The Primary Beam Component of the Axial Dose Profile $f_p(z)$

The aforementioned more complex beam model illustrated in Chapter 4 (Dixon et al. 2005) uses a modified convolution in lieu of the simpler first term of Eq. (6.1) to account for asymmetric penumbra produced by anode-tilt (at a target angle α) and includes the heel effect. For small angles, the heel effect is separable as a function $\rho(z)$; and, assuming a Gaussian focal spot, one obtains the analytic primary beam function derived in Chapter 4 (Dixon et al. 2005),

$$f_p(z) = \rho(z)\tilde{f}(z) = \rho(z)A_0\left\{\frac{1}{2}erf\left[\frac{\sqrt{\pi}}{C_L}\left(\frac{a}{2}+z\right)\right] + \frac{1}{2}erf\left[\frac{\sqrt{\pi}}{C_R}\left(\frac{a}{2}-z\right)\right]\right\} \quad (6.5)$$

where $erf(u)$ is the *error function* (Bracewell 2000); c_R and c_L and represent the focal spot penumbra at $z = \pm a/2$, respectively (the anode end is taken to be at $+z$ thus $c_L > c_R$), and $A_0 = f_p(0)$ in the case of MDCT. Also note that the heel effect has both an odd and an even (symmetric) component as shown in Appendix A. Assuming a rectangular focal spot intensity, likewise produces an analytical function (albeit less convenient) – namely an asymmetric trapezoid with a tilted roof.

If anode-tilt and the heel effect are ignored, Eq. (6.1) produces the same primary function as Eq. (6.5) which reduces to $c_L = c_R = c$ and $\rho(z) = 1$ in that case.

This primary beam function [Eq. (6.5)] produced near-perfect *matches* (not *fits*) to primary beam profiles measured using an IMRT film-digitizer system (Dixon et al. 2005) for a GE Lightspeed-16 scanner for both the 0.65 and 1.2 mm focal spots as has been amply illustrated in Chapter 4 (Dixon et al. 2005).

6.3.2 Derivation of the Scatter Component of the Axial Profile from the Scatter *LSF*

Having already obtained the primary beam component of $f(z) = f_p(z) + f_s(z)$ [given by Eq. (6.5)], the scatter component $f_s(z)$ is obtained from the convolution of the scatter $LSF = \eta \times lsf(z)$ with the primary beam core function $A_0 \Pi(z/a)$ as shown in Eq. (6.1), where $A_0 = f_p(0)$ for MDCT, and $lsf(z)$ is a unit area function [symbolically represented in terms of its width d as $lsf(z) = d^{-1}h(z/d)$, where η is the scatter-to-primary ratio (Dixon and Boone 2010; Dixon et al. 2005), namely,

$$f_s(z) = LSF(z) \otimes A_0 \Pi\left(\frac{z}{a}\right) = A_0 \eta \int_{-a/2}^{a/2} lsf(z - z')dz' \tag{6.6}$$

The scatter $LSF(z) = \eta \, lsf(z)$ based on a Monte Carlo simulation (Boone 2009) – dubbed the *dose spread function (DSF)* therein – was previously shown in Chapter 5 (Dixon and Boone 2010) to have the form of a double-exponential,

$$lsf(z) = (1 - \varepsilon)\frac{1}{d}\exp(-2|z|/d) + \varepsilon\frac{1}{\delta d}\exp(-2|z|/\delta d) \tag{6.7a}$$

with the parameter values in Eq. (6.7a) shown in Table 6.1, where η is the scatter-to-primary ratio.

On the *central axis*, the second term in Eq. (6.7a) is both small and transient (Dixon and Boone 2010) (having a small amplitude ε and a short relaxation length $\delta d \ll d$); and can be ignored, thus $lsf(z)$ can be approximated by its (re-normalized) asymptotic term as shown in Eq. (6.7b),

$$lsf(z) \cong d^{-1}h(z/d) \cong d^{-1}\exp(-2|z|/d) \tag{6.7b}$$

this approximation resulting in an error of less than 5% in the generated scatter function $f_s(z)$ (and less than 2% for $a > 40$ mm).

6.3.3 Calculation of the Complete Axial Dose Profile on the Phantom Central Axis

The scatter component is calculated using the simplifying approximation of Eq. (6.7b) in the derivation of $f_s(z)$, which facilitates the physical interpretation with no appreciable loss of accuracy [plus the final expression for $f_s(z)$ is easily modified by inspection to include

TABLE 6.1   Scatter LSF Parameters for 32 cm PMMA Body Phantom, 120 kVp, with Bow-Tie Filter

Phantom Axis	η	ε	d (mm)	δd (mm)
Central axis	13	0.015	117 mm	6.74 mm
Peripheral axis	1.5	0.304	88 mm	14 mm

Source: Dixon and Boone, *Medical Physics* (2011).

both terms in Eq. (6.7a) since both have the same functional form]. This more complex form (shown in Appendix B, Eq. B.2) is required on the peripheral axis where both ε and δd are larger.

Substituting the scatter *LSF* from Eq. (6.7b) into Eq. (6.6) and integrating the exponential form of $lsf(z)$ forces one to express $f_s(z)$ as two separate functions; $f_{si}(z)$ *inside the primary beam* region $|z| \leq a/2$; and $f_{so}(z)$ *outside the primary beam* $|z| \geq a/2$; with $A_0 = f_p(0)$, these functions being given by,

$$f_{si}(z) = A_0 \eta \left[1 - e^{-a/d} \cosh\left(\frac{2z}{d}\right) \right], \quad |z| \leq a/2 \tag{6.8a}$$

$$f_{so}(z) = A_0 \eta \sinh\left(\frac{a}{d}\right) \exp(-2|z|/d), \quad |z| \geq a/2 \tag{6.8b}$$

Despite its duality, the function $f_s(z)$ in Eqs (6.8a and 6.8b) is well-behaved with both $f_s(z)$ and its slope *exhibiting continuity across the boundaries* at $z = \pm a/2$; and $f_s(z)$ exhibiting (even) symmetry $[f_s(z) = f_s(-z)]$; and its smooth general shape is illustrated in Figure 6.4 for an $a = 80$ mm wide cone beam. It is noted from Eq. (6.8b) that the scatter tails of the profile, beyond the primary beam ($|z| \geq a/2$), have a pure exponential form $\exp(-2|z|/d)$ where $d = 117$ mm. Expressing this *lsf* in the alternate form $\exp(-\bar{\mu}z)$, results in the plausible value $\bar{\mu} = 2/d = 0.17\,\text{cm}^{-1}$.

The Complete Axial Profile on the Central Axis

Adding the primary beam component $f_p(z)$ from Eq. (6.5) to $f_s(z)$ from Eqs (6.8a and 6.8b) gives an analytic function for $f(z) = f_p(z) + f_s(z)$ on the central axis of the body phantom which can be plotted and compared to the experimental axial profile data (Mori et al. 2005) for beam widths (apertures) covering the range from $a = 28$ mm up to a cone beam width of $a = 138$ mm. We seek not only to *match* the shape of a given profile $f(z)$ but also to *match* the relative peak heights $f(0)_a$ with the theory for the gamut of available apertures (28 mm $\leq a \leq$ 138 mm).

6.4 COMPARISON OF THEORY WITH EXPERIMENTAL DATA

All the following experimental data is obtained at 120 kVp.

6.4.1 Primary Beam Function as Measured Free-In-Air

It has been previously demonstrated in Chapter 4 (Dixon et al. 2005) that the primary beam function given by Eq. (6.5) resulted in near-perfect *matches* (not *fits*) to primary beam profiles as measured free-in-air on a GE Lightspeed-16 scanner for a variety of apertures $a \leq 27$ mm, for both the 0.65 and 1.2 mm focal spot sizes.

A test of the appropriateness of Eq. (6.5) for cone beams was made using the widest beam ($a = 138$ mm) available (Mori et al. 2005) is shown in Figure 6.1 [with only the parameters F, z_a, and d_0 appropriately adjusted in accordance with the theory in Chapter 4 (Dixon et al. 2005) for the target angle and source-to-axis distance $F = 600$ mm of the 256 channel

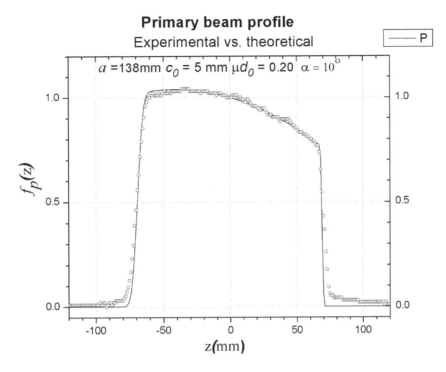

FIGURE 6.1 Match between experiment (Mori et al. 2005) and theory (Dixon et al. 2005) (solid line) described by Eq. (6.5) for the primary beam profile for the widest cone beam $a = 138$ mm. This is a *match* between a theory (having no adjustable parameters) and experiment – not a *fit* to experimental data.

(From Dixon and Boone, *Medical Physics*, 2011.)

system (see Appendix A)]; resulting in a good *match* between the theory and experiment as illustrated in Figure 6.1 (again, we re-emphasize that this represents a *match* – not a *fit* to the experimental data).

6.4.2 Dose Profiles $f(z)$ Measured on the Central Axis of the 32 cm PMMA Body Phantom Including Scatter

The Mori experimental data (Mori et al. 2005) for several representative apertures is plotted in Figure 6.2 together with the theoretical function $f(z)$, obtained by adding the primary component $f_p(z)$ from Eq. (6.5) to the scatter component $f_s(z)$ obtained from Eqs (6.8a and 6.8b), demonstrating a remarkably good *match* between theory and experiment for all beam widths ($a = 28$ mm to 138 mm), in spite of the fact that the theory allows no adjustment (it contains no empirical or adjustable parameters).

Agreement between the *peak height* function $f(0)_a \cong f_p(0)[1 + \eta(1 - e^{-a/d})]$ obtained from Eq. (6.8a), and the Mori experimental data (Mori et al. 2005) has been mathematically demonstrated in much greater detail in Chapter 5 (Dixon and Boone 2010). Note that the increase in the central peak dose $f(0)_a$ is due entirely to increasing scatter as beam width a increases; the primary beam contribution $f_p(0)$ remaining constant.

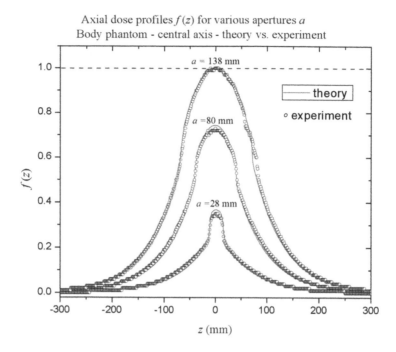

Axial dose profiles $f(z)$ for various apertures a
Body phantom - central axis - theory vs. experiment

FIGURE 6.2 Theory vs. experiment (Mori et al. 2005) on the central axis of the body phantom, normalized to unity for the widest ($a = 138$ mm) profile [on this scale the limiting (equilibrium) value of $f(0)_a$ for $a \geq 470$ mm is $A_{eq} = 1.4$].

(From Dixon and Boone, *Medical Physics*, 2011.)

Figure 6.3 shows the Figure 6.2 profiles *re-plotted using a logarithmic ordinate* in order to more clearly illustrate their purely exponential nature (as well as the accuracy of the match) in the scatter-tail region $|z| \geq a/2$; thus providing a clear visual confirmation of the robustness of our theory, which contains only parameters having a well-defined physical meaning.

Figure 6.4 shows the relative primary beam $f_p(z)$ and scatter $f_s(z)$ contributions to $f(z)$ for the $a = 80$ mm primary beam width to further illustrate the physics.

It is also noted that the parameters of the scatter $LSF = \eta \exp(-2|z|/d)$ are neither arbitrary nor adjustable – the values of the S/P ratio η and its equivalent width (Bracewell 2000) d, as well as its functional form, are all fixed by the physical assumptions (Boone 2009) implicit in the Monte Carlo simulation which begat the LSF, the validity of these assumptions (and the LSF itself) being confirmed by the congruence of theory and experiment shown in Figures 6.2 and 6.3. Abboud et al. have compared dose profiles from 10–300 mm (Abboud et al. 2010), also generated by a convolution kernel, to those generated by a complete MC simulation on the phantom central axis, obtaining good agreement between the respective profiles, apart from small differences in profile width and shoulders attributed to the neglect of beam divergence in the convolution model.

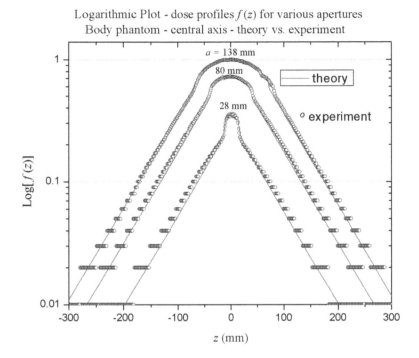

FIGURE 6.3 Semi-logarithmic plot of the data in Figure 6.2, illustrating the pure exponential nature of the scatter tails for $|z| \geq a/2$, as predicted by Eq. (6.8b). (The truncation of the raw experimental data in the low-signal extremities of the scatter tails results in the staircase appearance – *accentuated in this figure by the log scale*, but having little significance).

(From Dixon and Boone, *Medical Physics*, 2011.)

6.4.3 The Heel Effect and Wide Cone Beams

Despite the impressive tilt produced by the heel effect on the primary beam profile $f_p(z)$ for the widest $a = 138$ mm cone beam shown in Figure 6.1, there is little visual evidence of any heel effect remaining at depth *on the central-axis profiles* (Mori et al. 2005) shown in Figure 6.2 (the anode end is at $+z$). The increasing influence of the heel effect on $f_p(z)$, as the primary beam width a increases, is mitigated by the fact that the ratio of the scatter-to-primary contributions to $f(0)$ *increases* with a as $\eta(1-e^{-a/d})$, suppressing the relative primary contribution to $f(z)$ (and thence the heel effect) as a increases. [One expects a more pronounced heel effect on the peripheral axes due to the lower S/P ratio there ($\eta = 1.5$)].

Dose considerations notwithstanding, it should be noted that the heel effect can produce a noticeable noise gradient along z *in the image*, since the *primary photons* are the principal purveyors of image noise (Boone 2007).

The following additional considerations serve to limit the importance of the heel effect in quantitative CT dosimetry *based on the dose at $z = 0$*:

1. The relevant SCBCT dose [the dose on the central ray $f(0)$] was shown to be essentially unaffected by the heel effect.

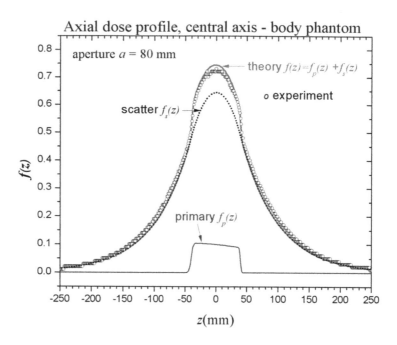

FIGURE 6.4 Illustration of the relative contributions of the theoretical scatter and primary functions to the total dose profile $f(z) = f_p(z) + f_s(z)$ for the $a = 80$ mm wide primary beam shown in Figure 6.3.

(From Dixon and Boone, *Medical Physics*, 2011.)

2. The conventional CT dose-descriptors $CTDI_L$ or $D_L(0) = p^{-1} CTDI_L$ in Eq. (6.4) are essentially impervious to the heel effect since the odd component of the heel effect cancels out in the symmetric integral of $f(z)$ over $(-L/2, L/2)$ – even further mitigated by the narrower fan beams typically utilized in conventional CT.

6.4.4 Stationary Phantom CT

Since $f(z)$ represents *the complete dose distribution* for SCBCT, there is no necessity to integrate it over $(-L/2, L/2)$ since CTDI does not apply as shown in Chapter 5 (Dixon and Boone 2010) [since A_{eq} is independent of aperture (beam width) it can be determined from D_{eq} (or $CTDI_\infty$) in the conventional CT mode using a single fan beam]. The integral form of the CTDI-based dose equations [Eqs (6.3 and 6.4)] was shown in Chapter 2 (Dixon and Boone 2010) to be the sole result of phantom translation over a length L, and thus CTDI cannot apply to SCBCT for which $L = 0$; indeed the integral format of Eqs (6.3 and 6.4) collapses, smoothly converging to $D_L(z) \rightarrow Nf(z)$ and $D_L(0) \rightarrow Nf(0)$ as table motion stops (in the limit as $b \rightarrow 0$, and $L = Nb \rightarrow 0$), in which case all N profiles (robbed of lateral dispersion) simply pile up on top of each other.

With SCBCT, one is not stitching together a series of profiles $f(z)$ as in axial or helical CT scanning – hence only $f(z)$ itself matters, and it is only the central peak dose $f(0)$ which is of primary importance - $f(0)$ corresponding directly to the CTDI-based dose $D_L(0) = p^{-1} CTDI_L$ in conventional CT, which likewise gives the dose at the center of the scan length ($z = 0$) as illustrated in Chapter 5.

6.4.5 Helical CT with Wide Cone Beams and with Table Translation – $CTDI_L$ Can Again Apply

Until recently, wide cone beams with $nT > 40$ mm were not typically used in conventional helical and axial CT using table translation (rather only for SCBCT); however, at least two systems now allow helical scanning using a wide ($nT = 80$ mm) beam, in which case one simply reverts to the conventional CTDI-based dose paradigm given by Eq. (6.4); however, the likely use of adaptive collimation (Deak et al. 2009) can perturb the central dose if only a few rotations are utilized.

6.5 DERIVING ANALYTICAL EQUATIONS FOR $CTDI_L$ AND RELATED QUANTITIES FOR CONVENTIONAL CT USING THE DOSE PROFILE FUNCTIONS PREVIOUSLY DERIVED ($L > \hat{a}$)

Performing the integration in Eq. (6.4) on the analytical profile function $f(z)$ allows us to obtain an analytic representation of $D_L(0)$ as well as $CTDI_L = pD_L(0)$ as a function of both scan length L and the z-aperture (primary beam $fwhm$) $\hat{a}$ of the integrand $f(z)$.

For any reasonable helical or axial scan series, the scan length L is larger than the primary beam width $\hat{a}$, thus the integral of the *primary component* $f_p(z)$ over $(-L/2, L/2)$ is the same as its infinite integral, being given by $A_0\hat{a} = f_p(0)\hat{a}$.

Adding to this the integral of the scatter component $f_s(z)$ in Eqs (6.8a and 6.8b) over the same interval $(-L/2, L/2)$ and dividing both integrals by the table increment b, Eq. (6.4) becomes,

$$D_L(0) = f_p(0)\frac{\hat{a}}{b}\left\{1 + \eta\left[1 - e^{-L/d}\frac{d}{\hat{a}}\sinh\left(\frac{\hat{a}}{d}\right)\right]\right\} \tag{6.9}$$

The equation for $CTDI_L$ is obtained by setting $b = nT$ (pitch $= 1$) in Eq. (6.9), namely,

$$CTDI_L = f_p(0)\frac{\hat{a}}{nT}\left\{1 + \eta\left[1 - e^{-L/d}\frac{d}{\hat{a}}\sinh\left(\frac{\hat{a}}{d}\right)\right]\right\} \tag{6.10}$$

$D_L(0)$ in Eq. (6.9) is seen to approach an equilibrium value for $L \gg d$ of $D_{eq} = (\hat{a}/b)f_p(0)(1+\eta) = (\hat{a}/b)A_{eq}$, thus D_{eq} is directly proportional to the aperture $\hat{a}$ as anticipated (Dixon et al. 2005) and shown in Chapter 5.

However, $D_L(0)$, thence $CTDI_L$ and also the relative approach to equilibrium function $H(L) = D_L(0)/D_{eq}$, also contain some additional dependence on $\hat{a}$ due to the factor $\frac{\sinh(\hat{a}/d)}{(\hat{a}/d)} = \frac{\sinh(x)}{x} \geq 1$, with $H(L)$ given by,

$$H(L) = \frac{1}{1+\eta} + \frac{\eta}{1+\eta}\left[1 - e^{-L/d}\frac{\sinh(\hat{a}/d)}{(\hat{a}/d)}\right] \tag{6.11}$$

Boone has shown (Boone 2007), using repeated Monte Carlo simulations for various apertures, that $CTDI_{100}$ ($L = 100$ mm) at 120 kVp exhibits a decrease of only 1.3% over the aperture range of 1–40 mm, which is in precise agreement with Eq. (6.10) which likewise

TABLE 6.2   The Coupling Factor Between Scan
Length L and Aperture $\hat{a}$

$\hat{a}$	$(d/\hat{a})\ \sinh(\hat{a}/d)$	L_{eq}
10 mm	1.001	468 mm
40 mm	1.02	470 mm
80 mm	1.08	477 mm
100 mm	1.13	N/A

Source: Dixon and Boone, *Medical Physics* (2011).

predicts a 1.3% decrease in $CTDI_{100}$ (from 0.605 $CTDI_\infty$ to 0.597 $CTDI_\infty$), and which also exhibits an absolute magnitude in close agreement with typical experimental values (Mori et al. 2005; Dixon and Ballard 2007) of $CTDI_{100} = 0.60\ CTDI_\infty$ (see Chapter 3).

However, *this small dependence on aperture $\hat{a}$ is of no significance for most axial or helical scan series* utilizing narrow fan beams with widths $\hat{a} \ll d = 117$ mm. That is, for beam widths up to $\hat{a} = 40$ mm for which $x = (\hat{a}/d) << 1$, the factor $x^{-1}\sinh(x) \approx (1 + x^2/6 + \ldots) \approx 1$, as also shown in Table 6.2. Thus, for typical fan beams widths ($nT \leq 40$ mm) used in axial or helical CT, Eqs (6.9–6.11) can be simplified by setting $x^{-1}\sinh(x) = 1$, as shown in Eqs (6.12–6.14), in which $p = b/nT$ denotes generalized pitch. It is also useful and informative at this point to include the equations for the peak axial dose $f(0)_a$ used for SCBCT for a side-by-side comparison as shown next ($\hat{a}$ is used for fan beam width to distinguish it from cone beam width a).

Conventional CT [$nT \leq 40$ mm] $L > \hat{a}$ **Stationary phantom CT (SCBCT)**

$$D_L(0) \cong \left(\frac{\hat{a}}{b}\right) f_p(0)\left[1 + \eta\left(1 - e^{-L/d}\right)\right] = p^{-1}CTDI_L \qquad f(0)_a \cong f_p(0)\left[1 + \eta\left(1 - e^{-a/d}\right)\right] \qquad (6.12)$$

$$CTDI_L \cong \left(\frac{\hat{a}}{nT}\right) f_p(0)\left[1 + \eta\left(1 - e^{-L/d}\right)\right] = pD_L(0) \qquad \text{Neither CTDI, nor } nT \text{ apply} \qquad (6.13)$$

$$D_{eq} = \left(\frac{\hat{a}}{b}\right) f_p(0)[1 + \eta] = \left(\frac{\hat{a}}{b}\right) A_{eq} \qquad A_{eq} = f_p(0)[1 + \eta] = \frac{1}{a}\int_{-\infty}^{\infty} f(z)dz \qquad (6.14)$$

$$H(L) \cong \frac{1}{1+\eta} + \frac{\eta}{1+\eta}\left(1 - e^{-L/d}\right) \qquad H(a) \cong \frac{1}{1+\eta} + \frac{\eta}{1+\eta}\left(1 - e^{-a/d}\right) \qquad (6.15)$$

$A_{eq} = (b/\hat{a})D_{eq}$ is a constant, independent of aperture $\hat{a}$ (and also nT) as previously demonstrated in Chapter 5 using the Mori data (Mori et al. 2005).

All of the aforementioned conventional CT equations assume $L = Nb > \hat{a}$ (a constant primary beam contribution) which is the typical case. [For helical scans using $nT = 80$ mm, $x^{-1}\sinh(x) \approx 1.1$, hence use of the more accurate Eqs (6.9–6.11) which retain the factor $x^{-1}\sinh(x)$ may be indicated in some cases]. The *free-in-air* equation for $CTDI_{air}$ can be obtained from Eq. (6.13) by setting $\eta = 0$ (no scatter).

6.5.1 Providing New Insight into the Physics of CT Dosimetry

The similarity between the conventional (helical and axial) and stationary phantom modalities illustrated by these equations is striking; however, it is noted that the conventional CT dose contains as an additional dependence on pitch, or specifically on $(\hat{a}/b)$ with $D_L(0)$ and D_{eq} both proportional to $(\hat{a}/b)$ – a logical result since the collimator aperture $\hat{a}$ controls the amount of primary beam energy allowed to impinge on the phantom ($\hat{a}$ acts as the energy gate), and $1/b$ determines the concentration of absorbed energy per unit phantom length, the ratio $\hat{a}/b$ representing the dose. There is, of course, no table increment for SCBCT in which only a matters.

These equations also suggest that a table increment of $b=\hat{a}$ is *special* – equalizing the conventional and SCBCT equilibrium doses $D_{eq}=A_{eq}$, as well as the peak doses $D_L(0)=f(0)_a$ at $z=0$ for the case $L=a$. In fact, for this *special* case ($b=\hat{a}=a/N$ and $L=Nb=a$), the SCBCT and the accumulated axial dose distributions are point-wise congruent $[f(z)_a=\widetilde{D}_N(z)]$ over all z [as previously demonstrated – both experimentally and theoretically in Chapter 5 (Dixon and Boone 2010)] – which is also eminently logical, since the same amount of energy is deposited over the same phantom length $L=a$. Since $\hat{a}$ represents the *fwhm* of the primary beam component, the *special* scan interval $b=\hat{a}$ is ideal, producing a naturally smooth accumulated axial dose distribution $\widetilde{D}_N(z)$ [Eq. (6.2)] on the central axis (in fact *perfectly smooth* apart from small oscillatory perturbations at the primary beam junctures due to the asymmetric penumbra and also the heel effect – both resulting from anode-tilt [as previously illustrated in Chapter 4 (Dixon et al. 2005) in a graphical simulation]. This *special* table increment ($b=\hat{a}$) also results in *perfect* (100%) dose efficiency corresponding to a *perfect pitch* $p=\hat{a}/nT>1$ which is equal to the over-beaming factor [producing *perfect smoothness* if anode-tilt is ignored (Dixon et al. 2005)].

6.5.1.1 The CTDI Equation – Looking Behind its Integral Facade

By stripping away the integral facade of the CTDI formula [as in Eq. (6.10) or Eq. (6.13)], we are able for the first time to observe its inner workings, from which the underlying physics can be deduced.

The basic physical form of all Eqs (6.12–6.15) (for conventional CT and SCBCT alike) consists of a constant primary-beam contribution $f_p(0)$ which is independent of both aperture a and scan length L, plus a scatter contribution to the dose at $z=0$ which increases as the length of phantom being irradiated increases, with scan length L in conventional CT (or with beam width a in SCBCT). This scatter increase ("field size dependence") being described by a common (and intuitively plausible) exponential growth function $f_s(0)=f_p(0)\,\eta(1-e^{-\lambda/d})$ for both modalities with $\lambda=a$ for SCBCT and $\lambda=L$ for conventional CT – which portends the eventual approach of the accumulated dose to an equilibrium (saturation) dose value for L (or a) $>4d=470$ mm – representing a probability of less than <2% (e^{-4}) that a photon scattered from the ends of the scan length at $z=\pm L/2$ (or beam width in SCBCT) can actually get back to $z=0$ to further augment the dose there. This is described by a common, relative approach to equilibrium function $H(\lambda)$ in Eq. (6.15), in which the first and second terms represent the relative primary and scatter contributions, respectively. A robust measure of machine "output" is $A_{eq}=f_p(0)[1+\eta]$, namely the *special* value of D_{eq} when $b=\hat{a}$.

The fundamental key to absolute dose common to both modalities is $f_p(0)$.

The *aperture-independent*, central-ray primary beam dose $f_p(0)$ at depth in the phantom is the fundamental "output" parameter common to all the preceding dose equations, *hence determining its value unlocks every equation* (6.9–6.15), for both conventional CT and SCBCT, allowing computation of the complete set of CT dose data for both modalities; namely the dose for any desired aperture, scan length, or pitch (including D_{eq} or A_{eq}); moreover, the value of $f_p(0)$ can be obtained from a *single measurement* of either $f(0)_a$ or $CTDI_L$ at a known aperture value a as shown in Chapter 5.

6.5.2 The Effect of the Weak Coupling Between Scan Length L and Beam Width $\hat{a}$ for Conventional CT Using Table Translation

As noted in the previous section, as beam widths $\hat{a}$ become comparable to $d = 117$ mm, the factor $(d/\hat{a})\sinh(\hat{a}/d) = x^{-1}\sinh(x)$ in Eqs (6.9–6.11) grows parabolically as $(1 + x^2/6 + \ldots)$, thus exhibiting a slow initial increase for small $x = \hat{a}/d$ (where $d = 117$ mm), as illustrated in Table 6.2.

This factor has the effect of slowing the approach to equilibrium [see Eqs (6.9 and 6.11)], its effect being larger the shorter the scan length L as shown by Eq. (6.9) in which the factor $\exp(-L/d)$ controls the approach [also illustrated by the $H(L)$ plots of Mori et al. (2005) for various apertures]; however, the effect of the factor $x^{-1}\sinh(x)$ on the equilibrium length itself L_{eq} is negligible as shown in Table 6.2 [where L_{eq} is defined as the scan length at which $D_L(0) \rightarrow (1 - e^{-4})D_{eq} = 0.982D_{eq}$ in Eq. (6.9), with $L_{eq} = 4d = 468$ mm for $\hat{a} \leq 10$ mm]; the effect of $x^{-1}\sinh(x)$ on the equilibrium dose value D_{eq} being nil as previously noted.

We should also note that for the CTDI-paradigm to apply at all, one requires several rotations (*at least* three) spread out over $L > \hat{a}$. For SCBCT, the aperture a itself is the only relevant variable contributing to an increase in scatter reaching $z = 0$, and $H(a)$ has no other dependence besides a (however, the equilibrium dose A_{eq} is not clinically relevant for SCBCT, since it is only attained for cone beam widths $a > 470$ mm which are not available on diagnostic SCBCT systems).

6.5.3 Application to Problems Beyond the Reach of the CTDI Method

6.5.3.1 Understanding and Exploiting the Symmetries Implied by the Convolution

If $f(z)$ from Eq. (6.1) is substituted into Eq. (6.3) for $D_L(z)$, one can see that $D_L(z)$ depends on the product of three convolutions $[LSF(z) \otimes A_0\Pi(z/\hat{a})] \otimes b^{-1}\Pi(z/L)$, and since the convolution operation is commutative and associative (Bracewell 2000), these can be performed in any order, e.g., $[LSF(z) \otimes \Pi(z/L)] \otimes \Pi(z/\hat{a})$, which implies a symmetry between $\hat{a}$ and L.

6.5.3.2 The Case of a Near-Stationary Phantom for which $L < \hat{a}$

This case represents a phantom translation less than the primary beam width, which necessarily implies a small pitch $p = b/nT < 1/N$ (since $L = Nb < \hat{a}$ and $nT < \hat{a}$), *such a protocol perhaps being clinically relevant to CT-fluoroscopy*. The IEC CT standards (IEC 2009) warn that CTDI does not apply for $L < nT$, but offers no alternative. The following derivation confirms the admonition and supplies the alternative.

The previous Eqs (6.9–6.15) applied to the usual axial or helical scan protocol with $L > \hat{a}$; however, the symmetry between L and $\hat{a}$ can be exploited to deduce the equation for $D_L(0)$ for the opposite case $L < nT < \hat{a}$, simply by swapping L and $\hat{a}$ in Eq. (6.9), and integrating the primary beam profile over a fraction $(L/\hat{a})$ of its width, with the result,

$$D_L(0) = Nf_p(0)\left\{1 + \eta\left[1 - e^{-\hat{a}/d}\frac{d}{L}\sinh\left(\frac{L}{d}\right)\right]\right\} \text{ for } L < \hat{a} \qquad (6.16)$$

Comparison with Eq. (6.9) *shows that the roles of L and â are reversed* such that $\exp(-\hat{a}/d)$ is now the controlling factor in growth of the accumulated dose at $z = 0$ due to scatter buildup. Comparison with Eq. (6.10) or Eq. (6.12) likewise confirms that $CTDI_L$ has no relevance in this domain (IEC 2009). In the ultimate limit as table motion stops $b \to 0$, $L = Nb \to 0$, (pitch $\to 0$), Eq. (6.16) approaches the limit $D_L(0) \to Nf(0)$ which represents all the profiles piling up on top of each other as $b \to 0$ noted previously. The basic equation Eq. (6.3) likewise converges directly to $D_L(z) \to Nf(z)$ in this limit [as does Eq. (6.2)].

For transitional cases where a and L are comparable, the symmetry between a and L is better illustrated by expressing the *cosh* and *sinh* functions used in Eq. (6.8) for $f_s(z)$ [and in Eqs (6.10–6.12)] in their exponential formats [$\frac{1}{2}(e^x \pm e^{\pm x})$]; thereby showcasing L and a in comparable functions and roles.

6.5.3.3 Evaluation of the Accumulated Dose $D_L(z)$ for an Arbitrary Value of $z \neq 0$

The CTDI, or any dose derived from it, is necessarily the dose at the center of the scan length $z = 0$; however, since the function $f(z)$ derived from Eq. (6.1) is readily integrable, one is no longer restricted to the dose at $z = 0$, rather one can integrate Eq. (6.3) to obtain $D_L(z)$ for any arbitrary z; however, various special solutions are simpler; such as calculating the dose beyond the scan interval $(-L/2, L/2)$ into the scatter tails $(z > L/2)$, which may have of practical utility in determining organ or fetal dose in that region. Also of interest is $D_L(\pm L/2)$, and readily calculable using these equations.

Dose calculation at a distance z_0 beyond the active scan length $z = L/2 + z_0$.

This problem is further simplified by using an additional minor constraint $z_0 > a/2$, thereby limiting the integration to the scatter function $f_{so}(z)$ in Eq. (6.8b). [This small gap can be filled in if desired by including half of $f_p(z)$ and $f_{si}(z)$ from Eq. (6.8a) in the integration, or by calculating $D_L(L/2)$].

The resulting integral for $z = L/2 + z_0$, with $z_0 > a/2$ results in,

$$D_L\left(L/2 + z_0\right) = \frac{1}{2}f_p(0)\left(\frac{a}{b}\right)\eta\left(x^{-1}\sinh x\right)\left(1 - e^{-2L/d}\right)\exp(-2z_0/d), \quad z_0 > a/2 \quad (6.17)$$

This relatively simple equation can replace a considerable amount of tabulated data (Boone et al. 2000) addressing the same problem, and can be normalized to $D_L(0)$ in Eq. (6.9) or to $CTDI_L$ in Eq. (6.10). Also see (Li et al. 2012) for a method using $H(L)$ described in Chapter 5.

6.6 THEORETICAL DOSE DISTRIBUTIONS ON THE PERIPHERAL AXES FOR SINGLE AND MULTIPLE ROTATIONS

On a peripheral axis, the parameters in the convolution equation [Eq. (6.1)] all vary (Dixon et al. 2005) with beam (gantry) angle θ, requiring a 2π integration over θ to obtain $f(z)$, which unfortunately results in the loss of its convolution format; however, it has been shown in Chapter 5 (Dixon et al. 2005) that the convolution can be restored to a good approximation by replacing the aperture $a(\theta)$ with its dose-weighted average value a', which represents the *fwhm* of the primary beam profile $f_p(z)$ on the peripheral axis, where a' differs only slightly from the minimum value of $a(\theta)$ at $\theta = 0$, namely $a' \approx 1.05a(0)$. This allows one to use a dose-weighted average of the scatter *LSF* in the convolution as was illustrated in some detail in Chapter 5.

6.6.1 An Analytical CT Dose Simulator – SIMDOSE

The derivation of the analytical primary beam profile in Eq. (6.5), and the required integration over θ for the peripheral axes, has previously been described in Chapter 4 for which an analytical CT dose simulator was also developed (Dixon et al. 2005), likewise using Eq. (6.5) for the primary beam component, but which also utilized an *empirical* scatter *LSF* (a double Gaussian). The angular integration and dose simulation was implemented in MATLAB® and its results demonstrated. It is clear that the empirical (double Gaussian) scatter *LSF* previously utilized cannot be generally correct, since the *actual LSF* is a double-exponential [Eq. (6.7a)]. However, simply by replacing the empirical scatter *LSF* with that of Eq. (6.7), which *LSF* begat the scatter function $f_s(z)$ in Eqs (B.1 and B.2) in Appendix B, bestows a new generality and a new name – SIMDOSE. The performance and utility of SIMDOSE is illustrated in Figure 6.5. The *match* to the experimental OSL data is excellent, and the effects of anode-tilt are readily evident (heel effect).

This peripheral axis simulation for $N = 11$ rotations run on a PC in MATLAB® required only 20 sec to generate both the helical and axial (not shown) dose distributions (despite the 2π integration over beam angle), which is considerably faster than a Monte Carlo simulation; noise free; and provides a level of detail (fine structure) superior to that of most MC simulations. Simulation on the phantom central axis is nearly instantaneous – no angular integration is required. Although this simulator contains an analytical *LSF* kernel derived from Monte Carlo, it is now free of this parentage, and is applicable to a wide variety of beam widths, scan lengths, pitch values, focal spots, etc. without requiring any new MC simulations.

6.6.2 Comparison of Theory with the Peripheral Axis Dose Profiles of Mori

For the GE fan beam data previously illustrated in Figure 6.5, the heel effect function $\rho(z)$ for a 7° target angle appeared successful on the peripheral axis; however, for the peripheral axis cone beam profiles measured by Mori (Mori et al. 2005), the heel effect produced only a modest tilt even for the widest cone beam ($a = 138$ mm), whereas the equation for $\rho(z)$ predicted a stronger heel effect. As previously shown in Chapter 4 (Dixon et al. 2005) the width of the primary beam component a' on the peripheral axis is only about 5% greater than its minimum projected value at $\theta = 0$, namely, $a_0 = a/M$ where $M = F/(F - 15 \text{ cm}) = 1.33$,

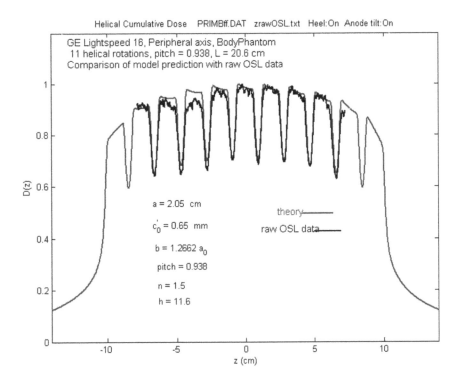

FIGURE 6.5 Simulation of the accumulated dose distribution on the *peripheral axis* of the 32 cm PMMA body phantom for a helical scan series of 11 rotations on a GE LS-16 CT scanner for a primary beam width (aperture) of $a = 2.06$ cm ($nT = 2.0$ cm), pitch of $p = 0.938$, resulting scan length $L = 20.6$ cm, small 0.65 mm focal spot as compared to the dose data measured using a Landauer OSL rod of 15 cm length for the same technique. (The measured OSL data is truncated at 15 cm).

(From Dixon and Boone, *Medical Physics*, 2011.)

due in large part to the bow-tie filter which has a significant effect in narrowing the width $a' = fwhm$ of the peripheral axis dose profiles (Dixon et al. 2005) $f(z)$, by "pinching off" contributions from larger beam angles θ, which correspond in turn to larger values of the projected aperture $a(\theta)$ on the peripheral axis.

Figure 6.6 shows the measured (body phantom) *peripheral axis* dose profile $f(z)$ for an aperture $a = 49$ mm from the Mori data (Mori et al. 2005), together with the corresponding SIMDOSE (Dixon et al. 2005) profile resulting from a *single axial rotation*. The *match* is clearly inferior to that of the central axis profiles – the theory *under-estimates* the scatter tails. Extraction of the simulated primary beam component (not shown) indicated the expected $fwhm$ $a' = 39$ mm $= 1.05a_0$.

In order to see if this scatter-tail divergence between the theory and the Mori data represents a consistent pattern, independent experimental profile data measured on a GE VCT scanner (using OSL strips recently published by Ruan et al. 2010) was utilized; the peripheral axis profile for $nT = 40$ mm at 120 kVp is plotted in Figure 6.7, together with the simulated profile. In contrast to the Mori peripheral axis data (Mori et al. 2005) in Figure 6.6, both the scatter tails as well as the robust heel effect are *well-matched* by the theory; however,

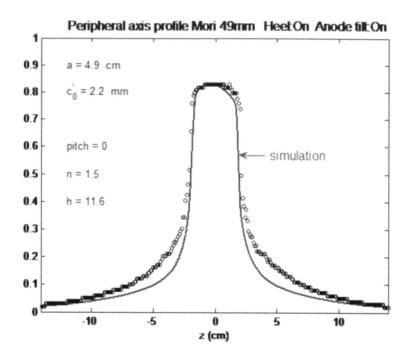

FIGURE 6.6 Simulated (solid curve) and measured (Mori et al. 2005) axial dose profiles on the peripheral axis for $a = 49$ mm.

(From Dixon and Boone, *Medical Physics*, 2011.)

the simulation *under-estimates* the experimental primary beam width (although the width $a' = 3.22$ cm $= 1.04\ a_0$ calculated by the simulation has the expected value). The aperture $a = 4.29$ cm corresponding to the large focal spot with $nT = 4$ cm is firmly established for the VCT scanner (Dixon and Ballard 2007) and thus unalterable according to our rules (no free-in-air primary beam profiles were measured hence no recourse is available). In contrast, the third peripheral axis data set (our own OSL data shown in Figure 6.5) was closely *matched* by the simulation on all counts (including beam width a').

The small discrepancies noted in Figures 6.6 and 6.7 are therefore not systematic, and the experimental data (acquired with relatively high-Z detectors) seem a more likely source [although both authors (Mori et al. 2005; Ruan et al. 2010) addressed the issue of energy response].

Although the *peripheral axis* match to the measured profile for wide cone beam is mediocre, this is not of great concern as long as the simulation predicts the relevant peak dose $f(0)$ with reasonable accuracy, which was shown to be the case in Chapter 5 for the peripheral axes.

Although application of the convolution theory to the peripheral axes involves a previously discussed approximation in Chapters 4 and 5 (use of an average scatter *LSF*) as well as some beam divergence issues as well as a reliance on a bow-tie/phantom primary transmission function (Dixon et al. 2005) $A_0(\theta)$, the theory and simulation can be judged as moderately successful on the peripheral axes.

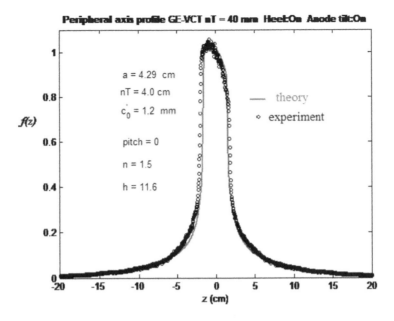

FIGURE 6.7 Axial dose profile peripheral axis – body phantom for GE VCT scanner $nT = 40$ mm, OSL data (Ruan et al. 2010).

(From Dixon and Boone, *Medical Physics*, 2011.)

6.7 APPLICATION TO *SHIFT-VARIANT* SCAN PROTOCOLS

The accumulated dose Eqs (6.2–6.4) for conventional helical or axial CT apply only in the case of identical dose profiles $f(z)$ spaced at equal intervals b along z, namely requiring *shift-invariance* with respect to z. The pencil chamber acquisition method likewise requires *shift-invariance* (all profiles must be identical to the one integrated by the pencil in order for the *predictive properties* of the CTDI-paradigm to apply); however, the direct "point-dose" measurement method using a small, Farmer-type ionization chamber (see Chapter 3) does not require *shift-invariance*.

Shift-invariance (and thus CTDI) therefore demands a constant aperture setting a (constant nT), constant mA and kVp, and constant pitch $p = b/nT$ (or a constant scan interval b).

The real hidden power of our method is freedom from the aforementioned constraints of *shift-invariance*, since each profile $f(z)$ can be individualized with respect to aperture $\hat{a}$, mA (and kVp), and spaced at irregular intervals $b(z)$, thereby allowing dose calculations for special scan protocols heretofore impossible using CTDI. This is easily accomplished for axial scans by generalizing Eq. (6.2) for $\tilde{D}_N(z)$– namely summing the profiles of variable aperture $a = a(k)$ and variable spacing $b = b(k)$, etc. For helical scans, one can approximate a continuously variable in a pitch $p = b/nT$ or aperture $\hat{a}$ as discretely stepped values on a rotation-by-rotation basis, using its axial surrogate $\overline{f}(z, b)_a$ in Eq. (6.2) $[\overline{f}(z, b)_a$ is the pseudo axial profile created by one helical rotation as defined in Chapters 3 and 4 which rigorously creates the actual helical dose distribution]. In fact, for the peripheral axes, SIMDOSE varies all parameters (Dixon et al. 2005) with $\theta = \omega t$, hence an arbitrary continuous variation of $\hat{a}(t)$ or $b(t)$ could be incorporated into the integral over θ for either the central or peripheral axes. An example would be a *helical* shuttle mode perfusion study in which the table oscillates back and forth using a variable pitch; or tube current modulation for which the profile amplitude (height) is varied. Actual rigorous analytic formulae are derived in Chapters 7 and 8 for shift-variant protocols including tube current modulation $i(z)$ and variable pitch $p(z)$ (e.g. a helical shuttle).

A beam profile resulting from asymmetric collimation can also be generated using the scatter *LSF*, and used to simulate helical protocols which asymmetrically reduce the aperture over the last ½ revolution or so to trim over-scanning. Many of these problems are now solvable using analytical simulation without resorting to Monte Carlo.

6.8 SUMMARY AND CONCLUSIONS

- The analytical equations describing the axial dose profiles $f(z)$ on the central axis of the 32 cm PMMA body phantom derived from a physical model (Dixon and Boone 2010; Dixon et al. 2005) –constrained by a rigorous adherence to the physics, thus containing no empirical functions or adjustable fit parameters – provided a good *match* to $f(z)$ in both shape and relative magnitude $f(0)$ when compared to the experimental profile data of Mori (Mori et al. 2005) over the range of apertures from $a = 28$ mm to $a = 138$ mm, as illustrated in Figures 6.2–6.4 [numerical matches to the peak dose data $f(0)_a$ having been thoroughly demonstrated and analyzed in Chapter 5]. This convolution-based model contains a previously derived analytical primary beam function in Chapter 4, accounting for both the focal spot length and its emission intensity distribution, as well as the asymmetric penumbra and heel effect produced by anode-tilt as a function of target angle. The scatter component of the beam profile is based on a scatter *LSF* kernel (DSF) derived from a Monte Carlo simulation, which, due to its simple form, can likewise be modeled analytically.

- This new scatter *LSF* kernel was incorporated into a previously developed (MATLAB-based) analytical beam simulator (replacing its empirical *LSF*) which provided a good match to complex peripheral axis helical dose distributions (Figure 6.5). Although this simulator contains an analytical *LSF* kernel derived from Monte Carlo, it is now free of this parentage and is applicable to a wide variety of beam widths, scan lengths, pitch values, focal spot sizes, etc. without requiring any new MC simulations.

- This formulation also makes it possible to integrate $f(z)$ analytically and thus to obtain for the first time analytical equations describing the accumulated dose $D_L(0)$ in Eq. (6.4) for conventional helical or axial scanning as well as for the $CTDI_L = pD_L(0)$ itself, the equations of which provided important (and definitive) analytical proofs including the weak coupling of scan length L and beam width $\hat{a}$ in conventional CT, as well as explaining other issues of importance for practical CT dosimetry.

- We likewise obtained a general function describing $D_L(z)$ for an arbitrary z by evaluating the convolution integral in Eq. (6.3); as illustrated for a special case for a dose-point at distance z_0 beyond the end of the scan interval $(-L/2, L/2)$, $z = L/2 + z_0$.

- The extended (as yet untapped) power of our method is freedom from the constraints of *shift-invariance* (Dixon et al. 2005) which limit the CTDI-paradigm to scan protocols for which each dose profile $f(z)$ maintains a constant collimator aperture $\hat{a}$, constant mA (and kVp), and spaced at equal intervals b (or constant pitch $p = b/nT$); whereas with our methodology, each dose profile can be individualized with respect to these parameters and spaced at irregular intervals $b(z)$ if desired. This also calls into question the meaning of $CTDI_{vol}$ presented for an auto mA protocol with $mA(z)$ modulation as will be described in Chapter 7. CTDI refers to the dose at $z = 0$ for multiple (identical) slices and thus CTDI cannot be modified on a slice-by-slice basis and then averaged or simply inserting the time-averaged value of a relevant parameter in the CTDI formula, e.g., in the case of variable $mA(z)$, the dose at $z = 0$ contributed by a given profile depends more strongly on its distance from the origin than its local mA, therefore the average mA is not a meaningful predictor of dose. In fact, Eqs (6.2–6.4), including the CTDI formula itself, *cannot be derived* (are invalid) for such *shift-variant* systems, hence *it is fallacious* to imagine a CTDI which is a function of time or z or to use a time-averaged value for any of the parameters therein such as a, b, p, and mA in these equations. In Chapter 7 a convolution model for tube current modulation is derived, and Chapter 8 extends this to a variety of other *shift-variant* scan protocols, such as variable pitch and aperture. Even dose calculations involving *shift-variant* phantoms can be handled once the dependence of the scatter *LSF* on phantom diameter has been established (a straightforward application of the same process previously used).

- The simple asymptotic form of the scatter LSF as a simple exponential literally begs a simple physical explanation, which we do not have.

ACKNOWLEDGMENTS

The author is grateful to **Dr. Shinichiro Mori** for supplying the numerical beam profile data for the various cone beam widths for the prototype 256 channel scanner (previously published in *Medical Physics*) which played a very important, central role in support of this work.

I would also like to thank **Dr. Chun Ruan** for supplying us with his beam profile data acquired using OSL strips on a GE VCT scanner (Ruan et al. 2010).

APPENDIX A: HEEL EFFECT FUNCTION

$\rho(z) \cong 1 - \langle \mu \rangle d_0 \dfrac{z}{z_\alpha}\left(1 + \dfrac{z}{z_\alpha}\right)$, where α = anode target angle $z_\alpha = F(\tan\alpha)$ = x-ray cutoff distance on anode end $(+z)$, where F = source to isocenter distance, $\langle \mu \rangle d_0 = \langle \mu \rangle \dfrac{d_e}{\tan\alpha}$ represents anode attenuation along the central ray of the x-ray beam (Dixon et al. 2005), $\langle \mu \rangle$ is the spectral average of $\mu(E)$ for tungsten, and d_e = mean depth of anode penetration for 120 keV electrons *along the electron beam axis* $[\langle \mu \rangle d_0 = 0.28$ for $\alpha = 7°]$.

APPENDIX B: SCATTER FUNCTION USING THE FULL DOUBLE-EXPONENTIAL *LSF* [EQ. (6.7A)]

As previously noted, the equations are easily modified by inspection, giving,

$$f_{si}(z) = A_0 \eta \left\{ \varepsilon \left[1 - e^{-a/d} \cosh\left(\frac{2z}{d}\right) \right] + (1-\varepsilon) \left[1 - e^{-a/\delta d} \cosh\left(\frac{2z}{\delta d}\right) \right] \right\}, \quad |z| \le a/2 \quad \text{(B.1)}$$

$$f_{so}(z) = A_0 \eta \left\{ (1-\varepsilon) \sinh\left(\frac{a}{d}\right) e^{-2|z|/d} + \varepsilon \sinh\left(\frac{a}{\delta d}\right) e^{-2|z|/\delta d} \right\}, \quad |z| \ge a/2 \quad \text{(B.2)}$$

GLOSSARY

SCBCT: stationary phantom cone beam CT

f(z): single rotation (axial) dose profile acquired with the phantom held stationary

b: table advance per rotation in conventional CT ($p = b/nT$ = generalized *pitch*)

a: the *z*-collimator aperture geometrically point-projected onto the axis of rotation (AOR); where $a \approx fwhm$ of the primary beam and where $a > nT$ for MDCT(over-beaming)

cone beam: typically having an aperture $a > 40$ mm

fan beam: a narrow beam ($nT \le 40$ mm)

â: denotes a fan beam aperture, to distinguish it from a cone beam aperture *a* in SCBCT

Deq: limiting value of accumulated dose at $z = 0$ approached to within e^{-4} (<2%) in conventional CT for scan lengths $L \ge Leq$

Leq: 470 mm on the central axis of the 32 cm diameter PMMA body phantom (Dixon and Boone 2010)

Aeq: $(b/a)Deq$ = *the equilibrium dose constant*, independent of aperture *a* (and thus *nT*)

η: scatter-to-primary ratio S/P (Dixon et al. 2005)

F: source-to-axis distance (focal spot to isocenter)

α: anode target angle

*z*α: $F(\tan\alpha)$ = anode x-ray cutoff distance on + *z* axis (anode end)

ρ(z): heel effect modulation function (Dixon et al. 2005) (see Appendix) $C_L = C_0[1 + a/2z_\alpha]$, $C_R = C_0[1 - a/2z_\alpha]$ penumbra widths at $z = -a/2$ and $z = a/2$, respectively (Dixon et al. 2003)

erf(z): error function. An analytic function (Bracewell 2000) given by the integral of a Gaussian over $(0, z)$

REFERENCES

AAPM, Report 111, Comprehensive methodology for the evaluation of radiation dose in x-ray computed tomography, American Association of Physicists in Medicine, College Park, MD, February (2010), http://www.aapm.org/pubs/reports/RPT_111.pdf.

Abboud S.F., Badal A.S., Stern H.S., and Kyprianou I.S., Designing a phantom for dose evaluation in multi-slice CT, SPIE Medical Imaging, (2010): Physics of Medical Imaging, E. Samei, N. J. Pelc, Editors, 7622-762232 [Proc. SPIE (2010)].

Barrett H.H., and Swindell W., *Radiological Imaging: The Theory of Image Formation, Detection, and Processing*, Revised ed., Academic Press, San Diego, (1981).

Bauhs J.A., Vrieze T.J., Primak A.N., Bruesewitz M.R., and McCollough C.J., CT dosimetry: Comparison of measurement techniques and devices. *Radiographics* 28, 245–253, (2008).

Boone J.M., Dose spread functions in computed tomography: A Monte Carlo study. *Med Phys* 36, 4547–4554, (2009).

Boone J.M., The trouble with $CTDI_{100}$. *Med Phys* 34, 1364, (2007).

Boone J.M., Cooper V., Nemzek W.R., McGahan J.P., and Seibert J.A., Monte Carlo assessment of computed tomography dose to tissue adjacent to the scanned volume. *Med Phys* 27, 2393–2407, (2000).

Bracewell R.N., *The Fourier Transform and Its Applications*, 3rd ed., McGraw Hill, Boston, (2000).

Deak P., Langner O., Lell M., and Kalendar W., Effects of adaptive section collimation on patient radiation dose in multisection spiral CT. *Radiology* 252, 140–147, (2009).

Dixon R.L., A new look at CT dose measurement: Beyond CTDI. *Med Phys* 30, 1272–1280, (2003).

Dixon R.L., and Ballard A.C., Experimental validation of a versatile system of CT dosimetry using a conventional ion chamber: Beyond $CTDI_{100}$. *Med Phys* 34(8), 3399–3413, (2007).

Dixon R.L., and Boone J.M. Analytical equations for CT dose profiles derived using a scatter kernel of Monte Carlo parentage with broad applicability to CT dosimetry problems. *Med Phys* 38, 4251–4264, (2011).

Dixon R.L., and Boone J.M., Cone beam CT dosimetry: A unified and self-consistent approach including all scan modalities—with or without phantom motion. *Med Phys* 37, 2703–2718, (2010).

Dixon R.L., Munley M.T., and Bayram E., An improved analytical model for CT dose simulation with a new look at the theory of CT dose. *Med Phys* 32, 3712–3728, (2005).

International Standard IEC 60601-2-44, *Medical Electrical Equipment — Part 2-44: Particular Requirements for the Basic Safety and Essential Performance of X-ray Equipment for Computed Tomography*, 3rd ed., International Electrotechnical Commission, Geneva, Switzerland, (2009).

Li X., Zhang D., and Liu B., Equations for CT dose calculations on axial lines based on the principle of symmetry. *Med Phys* 39, 5436–5352, (2012).

Mori S., Endo M., Nishizawa K., Tsunoo T., Aoyama T., Fujiwara H., and Murase K., Enlarged longitudinal dose profiles in cone-beam CT and the need for modified dosimetry. *Med Phys* 32, 1061–1069, (2005).

Ruan C., Yukihara E.J., Clouse W.J., Gasparian P., and Ahmad S., Determination of multislice computed tomography dose index (CTDI) using optically stimulated luminescence technology. *Med Phys* 37, 3560–3568, (2010).

Dose Equations for Tube Current Modulation in CT Scanning and the Interpretation of the Associated *CTDI*~vol~

7.1 INTRODUCTION

The CTDI-paradigm and associated equations are subject to strict physical constraints which are not widely appreciated (and often ignored) – the most important being the requirement for *shift-invariance* along the direction of table translation (z-axis).

The validity of the accumulated dose equations associated with the CTDI-paradigm (including the CTDI equation itself), *requires the existence of shift-invariant symmetry along z which imposes* a relatively strict set of constraints; namely, *identical in-phantom dose profiles f(z) spaced at equal intervals b along z* (as a result of phantom/table translation), and likewise demanding a *shift-invariant* phantom of uniform cross-section and density along the z-axis (but not necessarily cylindrical).

Shift-variant scan protocols include use of variable: pitch, z-collimator aperture $a(z)$, and the now-common tube current $i(z)$ modulation (TCM) for body scans; moreover, *shift-variance* renders Eqs (7.1–7.3) of the CTDI-paradigm unusable (including the CTDI equation itself).

New equations are derived which apply to variable i(z) TCM and which are immune to shift-variance. Equations which rigorously describe the auto TCM problem $i(z)$ are derived herein, in parallel with the equations of the CTDI-paradigm for fixed tube current i_0 (fixed mA) for a side- by-side comparison. Note that the variation of tube current with gantry angle does not result in *shift-variance*, hence we only consider z-axis $i(z)$ TCM, however, our equations are robust under simultaneous $i(z, \theta)$ TCM (Appendix A).

$CTDI_{vol}$ of the "first and second kinds." Despite the fact that CTDI, and thence $CTDI_{vol}$, do not apply to a *shift-variant* auto $i(z)$ TCM technique, there currently exists an imperative (IEC 2016) to report a value of $CTDI_{vol}$ for each and every patient exam – particularly in the case of the now-ubiquitous TCM protocols.

Therefore, the compromise required in applying the CTDI formula to the *shift-variant* auto TCM case with varying current $i(z)$ is *pre-ordained* (by physical law) to beget a flawed $CTDI_{vol}$, with a different interpretation and lacking the same physical significance.

In Chapter 6, analytical equations were developed representing axial dose profiles $f(z)$ which were further parlayed into a set of dose equations applicable to a gamut of problems (from wide cone beams to narrow fan beams for moving and stationary phantom alike), *including their proposed application to shift-variant systems, and which we deploy in this ever-widening shift-variant arena.*

7.2 METHODS

Rigorous analytical convolution equations describing the accumulated dose distributions $D_L(z)$ for both helical and axial scan trajectories are derived, allowing a side-by-side comparison of $D_L(z)$ for both $i(z)$ TCM and constant current i_0. Although these equations are complete, the analytic equations previously described in Chapter 6 are subsequently deployed to provide graphical dose simulations using various $i(z)$ distributions which more clearly illustrate the physical underpinnings of the accumulated dose. A glossary is appended for convenience.

7.2.1 Review of the Physical Meaning of the Traditional $CTDI_{vol}$ Based on Constant Tube Current (Constant *mA*)

For a scan technique using a constant tube current, $CTDI_{vol} = p^{-1}CTDI_w$ has a precise physical meaning: namely *the average dose across the area of the central scan plane* at $z = 0$ for a *pitch* $p = b/nT$, where b = the table advance per rotation (and $nT \equiv$ "N × T"). Since $CTDI_w$ depends inversely on nT, which subsequently cancels out in $CTDI_{vol} = p^{-1} CTDI_w$, the value of nT is moot, and $CTDI_{vol}$ is independent of nT, depending solely on the inverse b^{-1} of the table increment b. Note that $CTDI_{vol}$ is essentially an area average and is not a volume average (as might be inferred from its subscript), since no averaging over scan length L has been performed (measurement of $CTDI_{100}$ using a 100 mm long pencil chamber *does not imply any dose averaging over 100 mm*). The necessity for *shift-invariance* in CTDI can be operationally understood from the nature of a pencil chamber acquisition; namely, *all profiles in the scan series must be identical to the single profile integrated by the pencil chamber* in order for the equations and *predictive* nature of the CTDI-paradigm to apply. On the other hand, the direct measurement method using a small, Farmer-type ionization chamber (Dixon and Ballard 2007) described in Chapter 3 (and recommended by AAPM TG-111) is unaffected by *shift-variance*.

Integral dose E and *DLP* are typically immune to *shift-variance*.

$CTDI_L$ *is a group concept*, and refers to the dose at $z = 0$ *for an entire group of* (identical) *dose profiles* evenly spaced over the entire scan length L, with each profile augmenting the dose at the location of its neighbors via "scatter tails" extending well beyond the primary beam width a; these profiles collectively contributing scatter to build the cumulative dose distribution. By definition, $CTDI_L$ does not possess a value until the complete set of multiple profiles covering length L have been laid down, thus it is *fallacious* to consider a CTDI as a function of time t or z; i.e., $CTDI(z)$, or $CTDI_{vol}(z)$ are *absurdities*, as are "$CTDI_{vol}$ *per slice*" *or* "$CTDI_{vol}$ *per/sec*."

7.2.2 A Brief Review of the Relation of $CTDI_{vol}$ to Patient Dose

Since CTDI can only be measured in a *shift-invariant phantom* and thus can neither be measured in, nor directly applied to a (*shift-variant*) patient or humanoid phantom, then how can it be related to the dose received by an individual patient from a particular CT exam? Answer: The technique factors used in the patient exam are assumed to be applied to the same shift-invariant PMMA phantom, thereby allowing the calculation of a value for $CTDI_{vol}$ from the patient-based technique factors (apart from its basis of $CTDI_{100}$ which truncates $CTDI_{vol}$ to apply to a scan length of 100 mm). So, the 32 cm diameter PMMA body phantom serves in a dual role; as a *shift-invariant* measurement medium, and as a surrogate (*shift-invariant*) body habitus (a "body double") for purposes of assigning a "dose value" to every CT patient. $CTDI_{vol}$ also serves as a basis (or starting value) for dose estimates to the actual patient of improved accuracy; perhaps customized to their body size [using SSDE (AAPM 2011)]; corrected for actual scan length (Dixon and Ballard 2007), or used in the estimation of organ doses (Turner et al. 2010) or estimating fetal doses.

7.3 DERIVATION OF THE THEORETICAL EQUATIONS FOR AUTOMATIC TUBE CURRENT MODULATION (TCM)

The best way to unequivocally demonstrate the nature of the CTDI-paradigm, and to illustrate how it is affected by loss of *shift-invariance* (such as TCM), is to first derive the associated equations based on a *shift-invariant* fixed tube current i_0 (fixed *mA*) technique; and then repeat the derivation using a *shift-variant* $i(z)$ TCM technique; whereby these equations (via their differences) will provide a definitive and mathematically rigorous basis for comparison of the TCM technique to the CTDI-paradigm. The inapplicability of CTDI to *shift-variant* protocols *in no way precludes the possibility of a rigorous solution using basic physical principles*, whereas trying to force CTDI to work in such cases (or for a stationary phantom) will predictably lead to anomalies (Dixon and Boone 2010) as will be discussed in a later chapter.

7.3.1 *Shift-Invariant* Helical Technique Using Constant *mA*

Consider first a *shift-invariant* helical technique using constant *mA* in which no parameters vary with z (constant tube current i_0, pitch, aperture, etc.). It is relatively simple to derive the equations of the CTDI-paradigm for the accumulated dose $D_L(z)$ on the phantom central axis in this case as shown in Chapter 2. Translation of the table and phantom at velocity v produces a constant *dose rate* profile on the *phantom central axis* (Dixon 2003) in the form of a traveling wave $\dot{f}(z - vt) = \tau^{-1}f(z - vt)$ as depicted in Figure 7.1, where $f(z)$ is the single-rotation (axial) dose profile acquired with the phantom held stationary, and τ is the gantry rotation period (in sec); thus the dose accumulated at a fixed value of z (depicted

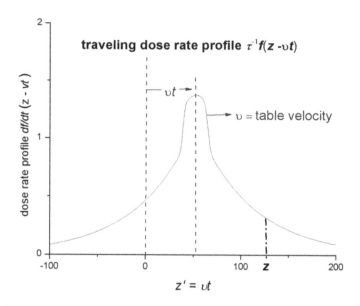

FIGURE 7.1   A traveling dose rate profile $\dot{f}(z - \upsilon t) = \tau^{-1} f(z - \upsilon t)$ in the phantom reference frame is created when an axial dose profile f(z) is translated along the phantom central axis z by table translation at velocity υ, where τ is the gantry rotation period (in sec), which has the familiar form of a traveling wave.

(From Dixon and Boone, *Medical Physics*, 2013.)

in Figure 7.1) as the profile travels by, is given by the time-integral of $\dot{f}(z - \upsilon t) = \tau^{-1} f(z - \upsilon t)$ over the total "beam-on" time t_0, namely,

$$D_L(z) = \tau^{-1} \int\limits_{-t_0/2}^{t_0/2} f(z - \upsilon t) dt = \frac{1}{b} \int\limits_{-L/2}^{L/2} f(z - z') dz' = \frac{1}{b} f(z) \otimes \Pi(z / L) \qquad (7.1)$$

the conversion from the temporal to the spatial domain having been made using $z' = \upsilon t$, scan length $L = \upsilon t_0$, and a table advance per rotation $b = \upsilon \tau$, resulting in the convolution equation [Eq. (7.1)] describing the total dose $D_L(z)$ accumulated at any given z-value during the complete scan.

Note that the aperture setting (and *primary beam fwhm*) for the dose profile shown Figure 7.1 is $a = 26$ mm, however, its wide scatter tails shown contribute to the dose at point z long before (and long after) the narrow primary beam has passed by, such that the *primary beam contribution* to $D_L(0)$ or to $CTDI_L$ is only a small fraction of the total dose. The accumulated dose at the center of the scan length $(-L/2, L/2)$ is easily obtained by setting $z = 0$ in Eq. (7.1), namely,

$$D_L(0) = \frac{1}{b} \int\limits_{-L/2}^{L/2} f(z') dz' \qquad (7.2)$$

The close resemblance of Eq. (7.2) to the CTDI equation is obvious; indeed, $CTDI_L$ is equal to $D_L(0)$ *for one particular value of table increment* $b = nT$ *(pitch $p = b/nT = 1$)*. A separate equation for $CTDI_L$ is therefore unnecessary and redundant, since $CTDI_L = pD_L(0)$. Nonetheless, we include it for later reference.

$$\text{CTDI}_L = \frac{1}{nT} \int_{-L/2}^{L/2} f(z')dz' \tag{7.3}$$

The approach to dose equilibrium function $H(L) = D_L(0)/D_{eq}$ also plays an important role in this paper, for which analytical equations have been derived in Chapter 6 (Dixon and Ballard 2007), and which can even be used to calculate $D_L(z)$ (as shown by Li et al. 2012).

The CTDI-paradigm therefore consists of Eqs (7.1–7.3), ranked in that order, however, the convolution Eq. (7.1) is sufficient and contains all the essentials (the complete set) from which Eqs (7.2 and 7.3) are readily obtained by setting $z = 0$. However, Eq. (7.3) is often taken as the starting point, using a "bottom-up" approach (and formula-plugging), which has led to considerable misunderstanding concerning the "CTDI equation" – further exacerbated by writing Eq. (7.3) as $CTDI_{100}$ with integration limits of ± 50 mm (Leitz et al. 1995). [Showing that $CTDI_{100} \rightarrow Nf(0)$ in the limit as table motion stops ($b \rightarrow 0$), is a good exercise for freeing one's mind of common misconceptions].

Note that $D_L(0) = p^{-1}CTDI_L$ represents the central or peripheral phantom axis component of a generalized $CTDI_{vol}$ (not restricted to $L = 100$ mm) upon application of the 1/3 or 2/3 weighting factor assigned to that axis (Leitz et al. 1995), which we will use in our analysis, since tube current $i(z)$ will be averaged over L (and not over 100 mm) in computing its average $\langle i \rangle$ as well as for $CTDI_{vol}$ (IEC 2016); further, we will principally use the phantom central axis component $D_L(0)$ as our $CTDI_{vol}$ surrogate to illustrate the physics without imposing excessive detail.

Eqs (7.1–7.3) likewise apply to helical scans on the peripheral phantom axis, as shown in Chapter 2 and to axial scans as well (also see Appendix A).

Thus Eqs (7.1–7.3) of the CTDI-paradigm apply to *all shift-invariant scans involving table translation (helical or axial)* on both the peripheral axes and on the central axis, as illustrated in Chapter 2.

7.3.2 Deriving the Dose Equations for a *Shift-Variant* TCM Technique

Auto *mA* TCM is one of the simplest *shift-variant* problems, since only the *amplitude* of the traveling dose rate profile of Figure 7.1 changes with tube current $i(z)$ (and not its shape) (Dixon and Boone 2013). For a variable $i(t)$, the average tube current $\langle i \rangle = \langle mA \rangle$ is taken over the entire scan time t_0, namely,

$$\langle i \rangle = \frac{1}{t_0} \int_{-t_0/2}^{t_0/2} i(t)dt = \frac{1}{L} \int_{-L/2}^{L/2} i(z')dz' \tag{7.4}$$

where $z' = \upsilon t$, $L = \upsilon t_0$, (assuming table velocity υ and pitch are constant). Also note that the total mAs = $\int i(t)dt = \langle i \rangle t_0$, where t_0 is the total "beam-on" time. In order to explicitly exhibit the effect of the variable tube current $i(t)$, it can be factored out from the traveling dose rate profile $\dot{f}(z - \upsilon t, t) = i(t)\hat{f}(z - \upsilon t)$, exhibiting an explicit dependence shown as the product of $i(t)$ with a current-independent shape function $\hat{f}(z)$, *namely the axial dose profile per unit mA.*

Additionally, since the traveling dose rate function at constant current $\dot{f}(z - \upsilon t) = \tau^{-1} f(z - \upsilon t)$ implicitly contains a constant current value i_0 as noted, the aforementioned dose rate profile is modified to the equivalent form,

$$\dot{f}(z - \upsilon t, t) = i(\upsilon t)\tau^{-1}\hat{f}(z - \upsilon t) = [i(\upsilon t)/i_0]\tau^{-1} f(z - \upsilon t)) \tag{7.5}$$

these profiles being related by $f(z - z') = i_0 \hat{f}(z - z')$ where $z' = \upsilon t$ as before, and $(i_0\tau)$ is the *mAs per rotation.*

Integrating the dose rate profile in Eq. (7.5) over the total scan time t_0, and converting to the spatial domain using $z' = \upsilon t$, $L = \upsilon t_0$, as before:

The modulated tube current (TCM) accumulated dose $\tilde{D}_L(z)$ is given by,

$$\tilde{D}_L(z) = \frac{1}{b} \int_{-L/2}^{L/2} i(z')\hat{f}(z - z')dz' = \frac{1}{b} \int_{-L/2}^{L/2} \left[\frac{i(z')}{i_0}\right] f(z - z')dz'$$

$$= \frac{1}{b}\hat{f}(z) \otimes [i(z)\Pi(z/L)] \tag{7.6a}$$

The accumulated dose for a constant tube current i_0 then follows from Eq. (7.6a),

$$D_L(z) = \frac{1}{b}i_0 \int_{-L/2}^{L/2} \hat{f}(z - z')dz' = \frac{1}{b} \int_{-L/2}^{L/2} f(z - z')dz' = \frac{1}{b}f(z) \otimes \Pi(z/L) \tag{7.6b}$$

where Eq. (7.6b) is identical to Eq. (7.1) for $D_L(z)$ in the CTDI-paradigm (likewise derived under *shift-invariant* conditions) at a constant current i_0, where i_0 is implicitly imbedded in $f(z)$. It will be useful later (for comparison purposes) to set $i_0 = \langle i \rangle$ of Eq. (7.4).

These equations (7.6a and 7.6b) are seen to be quite different – in the case of variable $i(z)$ one must convolve $\hat{f}(z)$ with the $i(z)$ function, or equivalently convolve $f(z)$ with the relative $i(z)$ function $[i(z)/i_0]$. The local TCM dose $\tilde{D}_L(z)$ at z is not proportional to the local $mA(z) = i(z)$ as is often erroneously stated, but rather depends on the current $i(z')$ over all z' locations due to scattered radiation which dominates the dose in CT. That is, the physical interpretation of the product (integrand) $i(z')\hat{f}(z - z')$ is that $i(z')$ determines the peak height of the profile at $z' = \upsilon t$, and $i(z')\hat{f}(z - z')$ represents the magnitude of the scatter tail at position z from that z'-profile, the contributions of which must be summed (integrated) *over every profile location z'.*

Therefore, the accumulated dose in CT at a given location z (including $CTDI_{vol}$ at location $z=0$) at a given time t depends on *mA past and mA yet to come*, and the scanner-reported "$CTDI_{vol}$ per slice" or $CTDI_{vol}(z)$ does not (and cannot) represent a dose, but merely represents the relative $i(z) \equiv mA(z)$. In fact, the dose $D_L(z)$ *even at constant mA is not constant* due to variations in such mutual profile scatter and peaks at z = 0, as seen in Chapter 2.

Both these equations (7.6a and 7.6b) show that neither the TCM dose $\tilde{D}_L(z)$ nor the constant current $D_L(z)$ at z are proportional to the local current $i(z) \equiv mA(z)$ at the same point. Therefore, *the average dose is not proportional to the average current, and likewise averaging the current is not the same as averaging the dose.* We can learn something about the physics involved from the convolution in Eq. (7.6a). First, the convolution (by its very nature) tends to smooth the tube current variations, reducing the effect of the local $i(z)$ on the accumulated dose $\tilde{D}_L(z)$ at the same point z, since $i(z)$ determines only the peak dose of a single profile located at z, but $\tilde{D}_L(z)$ depends on scatter contributions from many other profiles displaced relative to z. Physically, the wide scatter function $f_s(z)$ (Chapter 6) is primarily responsible for such smoothing, and scatter *cross-talk* between profiles reduces the influence of the local $i(z)$ on the dose $\tilde{D}_L(z)$ at that same point z.

Peripheral axes: Proof of the validity of Eqs (7.6a and 7.6b) for helical scans on the peripheral axes (using the angular average at a fixed z) is given in Appendix A; and their validity for axial scans [using the running mean (Dixon 2009; Bracewell 2000)] can be proven using the same methods utilized in the *shift-invariant* constant *mA* derivations by Dixon shown in Chapter 2 (Dixon 2003).

The constant tube current dose distribution is quite predictable – the convolution in Eq. (7.1) [or Eq. (7.6b)] $D_L(z) = b^{-1} f(z) \otimes \Pi(z/L)$ always produces a symmetric, "bell-shaped" distribution similar to Figure 7.2 having its maximum at $z=0$ as shown in Chapter 2; and $CTDI_{vol}$ at constant current (constant *mA*) and its single axis component $D_L(0) = p^{-1}CTDI_L$ both represent this peak dose at $z=0$, and this "predictability" adds value to the CTDI-paradigm (without which a single dose value like $CTDI_{vol}$ would have limited utility). Presuming that $CTDI_{vol}$ is computed using $CTDI_L$ instead of $CTDI_{100}$, but the *general shape* is the same.

Not so for TCM with variable tube current $i(z)$, for which the dose distribution $\tilde{D}_L(z)$ [Eq. (7.6a)] depends strongly on the exact functional form of $i(z)$, including the value and location of its peak dose (or peak doses), and information inside the dose domain is largely unpredictable using CTDI, as illustrated in the simulations to follow in Section 7.6.

The central doses at $z=0$, are shown for comparison as obtained by setting $z=0$ in Eqs (7.6a and 7.6b), namely,

for auto $i(z)$ TCM by Eq. (7.7a),

$$\tilde{D}_L(0) = \frac{1}{b} \int_{-L/2}^{L/2} \frac{i(z')}{i_0} f(z')dz' = \frac{1}{b} \int_{-L/2}^{L/2} i(z')\hat{f}(z')dz' \qquad (7.7a)$$

and for a *constant current* i_0 by Eq. (7.7 b) of the CTDI-paradigm,

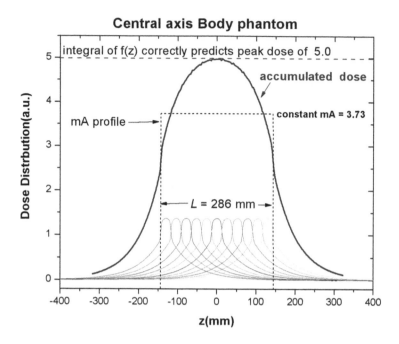

FIGURE 7.2 Accumulated dose at constant mA for a superposition (summation) of the 11 dose profiles depicted, each for an aperture of $a = 26$ mm and spaced at like intervals using a table increment $b = a = 26$ mm (no primary beam overlap). The peak accumulated dose at $z = 0$ contributed by the 11 adjacent profiles shown in Figure 7.2 exhibits a fourfold increase over the peak dose of a single axial profile due to scatter. In such a scan series with multiple adjacent profiles, the scatter contribution at $z = 0$ is built up by the scatter tails of the entire ensemble of profiles reaching back to $z = 0$. This curve is essentially congruent with the curve generated by the convolution in Eq. (7.1) as illustrated in Chapter 2, Figure 2.3.

(From Dixon and Boone, *Medical Physics*, 2013.)

$$D_L(0) = \frac{1}{b} \int_{-L/2}^{L/2} f(z')dz' = \frac{1}{b} i_0 \int_{L/2}^{L/2} \hat{f}(z')dz' \qquad (7.7b)$$

Eq. (7.7a) lacks the physical significance held by Eq. (7.7b), since the general asymmetry exhibited by $i(z)$ about $z = 0$ is likewise conveyed to $\tilde{D}_L(z)$ via Eq. (7.6a), thus $\tilde{D}_L(z)$ lacks the special symmetry about $z = 0$ always exhibited by the *shift-invariant* CTDI-paradigm [Eq. (7.7b)] by virtue of its constant current i_0.

Our derivation of the equations for the auto TCM mode [Eq. (7.6a)] did not require the imposition of *shift-invariance*, and is therefore quite general, robust, and relatively easy to apply – a straightforward convolution of $\hat{f}(z)$ with $i(z)$ as described by Eq. (7.6a) produces the complete TCM accumulated dose $\tilde{D}_L(z)$ – so we could likely define a more meaningful dose index for the TCM mode; thereby avoiding the inevitable pitfalls encountered *by forcing a (shift-invariant) CTDI to operate in a shift-variant territory*. However, a modified "CTDI$_{vol}$ of the second kind" has already preceded us (IEC 2016), which is different from CTDI$_{vol}$ for a constant current i_0; these differences (and their significance) have not been previously described (and have seemingly been ignored). This analysis is best temporarily postponed and presented in conjunction with the TCM simulations to follow.

7.4 TOTAL ENERGY ABSORBED FOR AUTOMATIC TUBE CURRENT MODULATION (TCM)

Our evaluations of dose distributions $D_L(z)$ and specific point-dose values (such as the central dose $D_L(0)$ at $z = 0$) have failed to provide any useful connection between "$CTDI_{vol}$ of the first and second kinds," however, the problem has dropped an occasional hint that *total mAs* $= \langle i \rangle t_0 = i_0 t_0$ and thence total energy E absorbed in the phantom (integral dose) and its surrogate DLP are the significant common quantities. The convolution equations derived for $\tilde{D}_L(z)$ in Eq. (7.6a) for auto $i(z)$ TCM, and for $D_L(z)$ in Eq. (7.6b) for constant current i_0 immediately suggest that (integral dose) E will provide the connection between TCM and constant tube current protocols in our search for commonality and physical meaning.

The total energy absorbed in the phantom along (and about) a given phantom z-axis can be calculated using the infinite integral $E = \int\limits_{-\infty}^{\infty} D_L(z)dz$, and since the infinite integral of a convolution is separable (Bracewell 2000) as $\int\limits_{-\infty}^{\infty} g(z) \otimes h(z)dz = \int\limits_{-\infty}^{\infty} g(z)dz \int\limits_{-\infty}^{\infty} h(z)dz$, evaluation of the energy integral becomes trivial when applied to Eq. (7.6a) for variable $i(z)$ and Eq. (7.6b) at constant tube current i_0, and it is readily shown that *the total energy absorbed in both cases is the same (equal)* and given by,

$$E = Lb^{-1} \int\limits_{-\infty}^{\infty} f(z)dz = N \int\limits_{-\infty}^{\infty} f(z)dz = LD_{eq} \tag{7.8}$$

Note also that the DLP formula has the same format as Eq. (7.8) and thus maintains a proportionality to E, namely $DLP = H(100)E = 0.61E$ on the central axis (see Chapter 6), hence all our remarks concerning the integral dose E also apply to DLP (see Chapter 2 for a more complete treatment of E).

But neither E nor DLP depend on the scan length L per se, so the form of Eq. (7.8) can be misleading since hidden variables therein (when explicitly exposed) reveal the actual physical dependences of E (and DLP). These hidden variables are $L = Nb$ and the b^{-1} dependence of D_{eq} (and $CTDI_{vol}$), with the result that the table increment b cancels out in the product of LD_{eq} in Eq. (7.8) (and likewise in $DLP = L \times CTDI_{vol}$), such that E and DLP depend only on the number of rotations N and the average *mAs per rotation* $\langle i \rangle \tau$, or simply on their product $\langle i \rangle N\tau = \langle i \rangle t_0 = $ *the total mAs* delivered during the "beam-on" time t_0 (assuming constant kV and bow-tie filter/FOV); apart from an additional direct-dependence on the z-collimator aperture a (which acts as an energy gate) and which is hiding in every *dose* and CTDI equation (its role being explicitly revealed in Eqs (6.12–6.15) of Chapter 6 in which the integral facade of the CTDI-paradigm [Eqs (7.2 and 7.3)] has been stripped away).

In summary, for a given kV and beam filter, the total energy absorbed (integral dose) E *(and its surrogate DLP) depend only on the product of total mAs and z-collimator aperture a.* So that's it – E and DLP depend only on total mAs $= \int i(t)dt = \langle i \rangle t_0$, and collimator aperture a, and are indifferent as to how the N rotations are spread out along the z-axis; in fact, E and DLP remain unchanged even if the table should stop moving ($b=0$ and $L=Nb=0$). However, for a given E, *the accumulated dose* $D_L(z)$ *will depend profoundly on how the energy* E *is spatially distributed along* z (on E per unit length); depending on L and the functional form of $i(z)$ for a TCM protocol. But E/L does not represent the average dose over the scan length, since some of the energy is deposited by scatter beyond the scan interval as shown in Table 7.1, in which E_{in}/E is the fraction of the total energy E deposited inside the scan interval $(-L/2, L/2)$ – details shown in Section 7.7.

7.5 THE TROUBLE WITH THE REPORTED VALUES OF $CTDI_{\text{VOL}}$

7.5.1 Defining $CTDI_{\text{vol}}^{\text{TCM}}$ for a Shift-Variant, Auto TCM Protocol

The ad hoc method implemented by CT manufacturers (via the IEC 2016) is made using a circuitous and curious argument based on a $CTDI_{\text{vol}}(t)$ [or $CTDI_{\text{vol}}(z)$]. Suffice it to say that this argument is the equivalent of computing the average tube current $\langle i \rangle = \langle \text{mA} \rangle$ over the *total scan time* t_0 (or total scan length L) as shown in Eq. (7.4), and then plugging $\langle i \rangle$ into Eq. (7.7b) of the CTDI-paradigm, under the pretext that $\langle i \rangle$ represents a *bona-fide constant current value*; thus creating an *illusory shift-invariance* which seemingly "allows" use of the CTDI-paradigm to calculate the dose (which also ignores the inconvenient fact that $CTDI_{\text{vol}}$ is based on a scan length of 100 mm whereas the average current is computed *for the entire scan length* L). On a simpler (more straightforward) level, this amounts to replacing $i(z)$ inside the integral in Eq. (7.6a) for $\tilde{D}_L(z)$ by its average value $\langle i \rangle$ and pulling it outside the integral as a constant current $i_0 = \langle i \rangle$, thus Eq. (7.6a) → Eq. (7.6b) of the CTDI-paradigm with $CTDI_{\text{vol}}^{\text{TCM}} = D_L(0)$ given by Eq. (7.7b). Replacing $i(z)$ with by its average $\langle i \rangle$ is by no means equivalent to averaging the

TABLE 7.1 Various Accumulated Dose and Energy Deposition Fractions for the Central Axis of the Body Phantom at 120 kV and Constant mA (Aperture a $\ll$ L) for which d = 117 mm and the S/P Ratio $\eta = 13$, as Calculated from Eqs (7.12–7.17)

Scan Length L (mm)	$H(L) = \dfrac{D_L(0)}{D_{eq}}$	$x = L/d$	$\dfrac{E_{in}}{E} = \dfrac{\bar{D}_L}{D_{eq}}$	$\dfrac{D_L(\pm L/2)}{D_L(0)}$	$\dfrac{\bar{D}_L}{D_L(0)}$	$\dfrac{\bar{D}_L}{D_{100}(0)}$
58.5 mm	0.43	0.5	0.41	0.75	0.95	0.66
100	0.60	0.855	0.56	0.69	0.92	0.92
200	0.83	1.71	0.76	0.58	0.89	1.25
286	**0.92**	**2.44**	**0.81**	**0.59**	**0.88**	**1.33**
470	0.985	4.0	0.88	0.51	0.90	1.44
702	0.995	6.0	0.92	0.50	0.93	1.51
∞	1.00	∞	–	0.500	1.00	1.64

Source: Dixon and Boone, *Medical Physics* (2013).
The bold values represent our example in Figure 7.6

dose, since neither $\tilde{D}_L(z)$ nor $D_L(z)$ are proportional to $i(z)$; *not even at constant current* $i(z) = i_0$ (see Figure 7.2). The mathematical formula for the IEC version of $CTDI_{vol}^{TCM}$ is shown in Section 7.7.1, Eq. (7.18).

The average tube current $\langle i \rangle$ which replaced $i(z)$ is not an actual constant tube current but rather a mathematical construct. It does not create the smooth bell-shaped dose distribution having a peak value equal to the dose $D_L(0)$ at $z = 0$ (Figure 7.2) associated with constant current. Basically, the dose distribution created by this mathematical average current $\langle i \rangle$ is purely a fantasy – it does not exist. The pretext that $\langle i \rangle$ behaves like a real constant current i_0 amounts to simply throwing a cloak over $i(z)$, hiding the reality that $i(z)$ is still underneath, working and varying, to create an actual dose distribution $\tilde{D}_L(z)$ significantly different from the *fantasy distribution* just described, *having a peak value that is neither equal to the reported* $CTDI_{vol}^{TCM}$ *nor likely to occur at* $z = 0$, *as will be illustrated later in a variety of simulations.* But unfortunately, $CTDI_{vol}^{TCM}$ is not a fantasy, but it is actual scanner-reported value of $CTDI_{vol}$ for a TCM scan (IEC 2016), despite the use of the constant current $\langle i \rangle$ subterfuge to the extent that there is no recognizable connection between $CTDI_{vol}^{TCM}$ and the real dose distribution $\tilde{D}_L(z)$ for the TCM technique it is supposed to characterize. There exists an actual (complex) TCM dose distribution $\tilde{D}_L(z)$ generated by $i(z)$, about which $CTDI_{vol}^{TCM}$ has little or no useful information to offer – it offers only a description of an implied (but imaginary) *dose distribution based on an average constant current.* The complexity of $\tilde{D}_L(z)$ for TCM makes it difficult to characterize with a single dose number such as $CTDI_{vol}$, however, we have derived easily implemented equations for auto TCM which can be used to give the complete $\tilde{D}_L(z)$.

7.5.2 Tube Current Modulation (TCM) Versus Constant Current Summary

Integral dose E and its surrogate DLP are robust between the two tube current modalities.

Since the integral dose E and the DLP depend only on the product of *total mAs* and aperture a; using any technique having the same total mAs $= \int i(t)dt = \langle i \rangle t_0$ guarantees that the total energy deposited E (and DLP) will be the same, whether for a TCM or manual (fixed mA) technique with $i_0 = \langle i \rangle$.

Neither dose nor $CTDI_{vol}$ are robust between the two tube current modalities.

Dose and $CTDI_{vol}$ are not robust, and $CTDI_{vol}^{TCM}$ has no predictive power. *The scanner-reported value of* $CTDI_{vol}^{TCM}$ *is unrelated to the either the dose at* $z = 0$ *or to the peak dose.* Using the pretext that the average tube current $\langle i \rangle$ in TCM is equivalent to a constant current $i_0 = \langle i \rangle$, we must avoid being duped into picturing a *fantasy (bell-shaped) dose distribution* which has a peak dose at $z = 0$ equal to the reported $CTDI_{vol}^{TCM}$ but which has no connection to the reality of the dose distribution created by the actual tube current $i(z)$ as will be illustrated by our simulations.

Computation of the convolution for auto mA (TCM) in Eq. (7.6a) is easily implemented using commonly available computer algorithms (e.g. MATLAB®, EXCEL, or even some plotting software); and for those familiar with the convolution method, it affords a qualitative simulation (Bracewell 2009) of $\tilde{D}_L(z)$ which the trained eye can visualize for any $i(z)$ variation; however, the methodology in the following section provides an exact graphical simulation and also a greater physical insight (and perhaps a welcome respite from the foregoing mathematical analysis).

Having finished the formal mathematical derivation, we will revert to the common vernacular in which tube current i is simply referred to as mA (likewise tube potential as kV), so $i(z) \equiv mA(z)$ and $\langle i \rangle \equiv \langle mA \rangle$ = average mA as in Eq. (7.4).

7.5.2.1 A Pencil Chamber Measurement Has No Utility Whatsoever for Auto TCM nor for Shift-Variant Techniques in General

Consider the fallacy of making a pencil chamber measurement for every $mA(z)$ in a TCM scan to get $CTDI_{100}(z)$ and then averaging $CTDI_{100}(z)$. But this so-called "$CTDI_{100}(z)$" is still the dose at the center of a 100 mm scan length $z=0$, and is not the same as the dose $D_L(z)$ over the scan length (which varies significantly over L, even at constant mA). This is essentially the IEC TCM method – IEC 2016; which is simply equivalent to averaging $mA(z)$ [see Eq. (7.18)]. The reported "$CTDI_{vol}$ per slice" is not the *local dose* at z but rather represents *local mAs* per slice (per rotation). Eq. (7.6a) shows that the TCM dose $\tilde{D}_{100}(z)$ is not proportional to $i(z) = mA(z)$, but rather to a convolution of the dose profile with $i(z)$.

7.6 A GENERAL METHOD FOR HANDLING *SHIFT-VARIANT* PROTOCOLS

In this section, we will deploy our analytical dose profile equations (Dixon and Boone 2011) $f(z) = f_p(z) + f_s(z)$ illustrated in Chapter 6 to simulate the dose distributions resulting from several $mA(z)$ profiles simulating auto TCM techniques, plotted in a series of figures to more clearly illustrate the underlying physics, as well as illustrating a novel and flexible methodology which can be used to simulate other more complex (*shift-variant*) scan protocols which are not as readily amenable to the derivation of descriptive equations such as those previously derived herein for the *shift-variant* TCM problem (equations variable for aperture and pitch are illustrated in Chapter 8).

The analytical equation describing the primary beam component $f_p(z)$ was previously derived in Chapter 4 from first principles (Dixon et al. 2005) and the analytical equation for the scatter component $f_s(z)$ was obtained by convolving the primary beam core function with an analytical scatter kernel $LSF(z)$ of Monte Carlo parentage (Dixon and Boone 2010; Boone 2009) – these being illustrated in Chapter 6. The *match* between these theoretical equations and experimental data (Mori et al. 2005) has previously been demonstrated in Chapter 6 (Dixon and Boone 2011).

The accumulated dose distribution resulting from N axial profiles $f(z)_a = i(z)\hat{f}(z)_a$ of variable aperture a, and spaced at arbitrary intervals b along z as desired, can be obtained using the same summation as for an axial scan series (Dixon 2003),

$$\tilde{D}_N(z) = \sum_{k=-J}^{J} f(z-kb)_a = \sum_{k=-J}^{J} mA_k \hat{f}(z-kb_k)_{a_k} \qquad (7.9)$$

Where k denotes the k_{th} rotation of a total of $N = 2J + 1$ rotations, $\hat{f}(z)$ denotes the axial profile per unit mA as before, and a_k and b_k denote possible changes in aperture a (primary beam *fwhm*) and scan interval b for handling *shift-variant* scan techniques on a rotation-by-rotation basis, in which the mA is also assumed to vary step-wise with each rotation $i_k = mA_k$ (other variables such as kV may also be included if desired). [Dixon et al. have previously shown (Dixon et al. 2005) how a helical scan series can likewise be rigorously cast into the same axial summation format as illustrated in Chapter 4.] We will deploy the convolution equations in the simulations of Chapter 8.

Since our aim is to gain a physical understanding of the dose distribution for TCM (auto $mA(z)$ scans), we will restrict our treatment to the *central axis of the 32 cm diameter PMMA "body" phantom* at 120 kV, which will provide the understanding we seek and simplifies the derivation (extending it to the peripheral axis is straightforward, but the equations are a bit more complex). Thus, for a constant mA, $D_L(0) = p^{-1}CTDI_L$ is our surrogate for the central axis component of $CTDI_{vol}$ [the actual component is $1/3\ D_L(0)$ but we needn't bother with the 1/3]. We must also be cognizant of the fact that the actual scanner-reported $CTDI_{vol}$ is based on $CTDI_{100}$, thus our comparisons using the surrogates $CTDI_L$ and $D_L(0) = p^{-1}CTDI_L$ (albeit more logical) should be applied only when comparing dose distributions for a fixed scan length (as in the graphical simulations to follow). When comparing doses to scanner-reported doses for variable scan lengths, we must revert to $D_{100}(0) = p^{-1}CTDI_{100}$ (easily related at any time to scan length L using $D_L(0) = D_{100}(0)$ $[H(L)/H(100)]$ and likewise for converting $CTDI_L$ to $CTDI_{100}$).

Our simulations will use the theoretical beam profile based on our analytical equations in Chapter 6 and shown in Figure 7.2 for an aperture of $a = 26$ mm, using a like table increment (scan interval) of $b = a = 26$ mm, with $N = 11$ rotations to represent a realistic clinical scan length of $L = Nb = 286$ mm [but short of the $N = 19$ rotations ($L = Nb = 494$ mm) required to achieve dose equilibrium at $z = 0$].

7.6.1 Simulation for a *Shift-Invariant* Constant mA

The first example shown is for a constant mA to establish a familiar baseline as illustrated in Figure 7.2.

Integrating any one of the identical analytical dose profiles $f(z)$ shown in Figure 7.2 using Eq. (7.2) of the CTDI-paradigm, predicts precisely the same value for the maximum central dose as is obtained from the numerical summation of profiles in Eq. (7.9), namely,

$$D_L(0) = \frac{1}{b} \int_{-143\,mm}^{143\,mm} f(z)dz = 5.0.$$ Likewise, the dose distribution $D_L(z)$ shown can readily be

generated using the convolution in Eq. (7.1) and illustrated in Figure 7.2 and in Chapter 2, Figure 2.3. The limiting equilibrium dose $D_{eq} = 5.41$ is also obtained by direct integration (extending the above integral to "infinity"). Note that "$CTDI_{vol}$ per slice" or $CTDI_{vol}(z)$ (*oxymorons*) would be constant in this case, implying a dose profile having *the same rectangular shape as the mA profile* of height $CTDI_{vol} = 5.0$.

7.6.2 Simulation of *Shift-Variant* Protocols (such as *z*-axis TCM) for which the Formulae of the CTDI-Paradigm Eqs (7.1–7.3) are Not Valid

The modified convolution equation [Eq. (7.6a)] derived for TCM with auto $i(z) = mA(z)$ is demonstrably quite different from Eqs (7.1–7.3) of the CTDI-paradigm, offering a guarantee that the CTDI-based dose equations will not work; moreover, the simple expedient of "plugging" Eqs (7.2 or 7.3) with the average mA is likewise invalid. However, integral dose E and DLP were shown to be robust for the same total mAs $= \langle mA \rangle t_0 = \int mA(t)dt$.

1. *The misconception that a strong coupling exists between the local mA(z) and the local accumulated dose $\tilde{D}_L(z)$ has somehow gained traction* (having recently been stated in a variety of venues with increasing frequency), which unfounded assumption (myth) will be refuted in the examples that follow. In fact, inspection of Figure 7.2 illustrates that even when the mA is constant over the entire scan interval $(-L/2, L/2)$, it fails to "lock-in" the accumulated dose $D_L(z)$ to a constant value, that is $D_L(z)$ varies from its peak value of 5.0 at $z=0$ down to about 2.5 at $z=\pm L/2$. It is the long reach of the scatter tails from non-local dose profiles, which represent a significant contribution to the local dose, and which foil the myth. This misconception is particularly disturbing when stated as "$CTDI_{vol}(z)$ is proportional to the local $mA(z)$"; which isn't true – even at constant mA. The CTDI gives the dose at $z=0$ for *an entire ensemble of identical dose profiles*, such as illustrated in Figure 7.2.

2. Specific dose simulations with variable $mA(z)$ as with TCM.

A clinical example of a TCM $mA(z)$ function for a scan covering both chest and abdomen is shown in Figure 7.3 as a point of reference.

The family of $mA(z)$ profiles used in our simulations:

We have specifically constructed a family of three different (somewhat arbitrary) auto $mA(z)$ profiles designed to demonstrate the physics involved, plus a fourth *constant mA* profile for comparison. These have not been designed to emulate TCM mA variations for any specific anatomical region, however, the dynamic range of the relative mA variations (a factor of 6) is within the realm of observed clinical mA variations. For convenience, the mA steps have integer values from 1 to 6, which can be considered to be in units of 100 mA (100–600 mA).

All mA(z) profiles were specifically designed to have the same average mA over the same scan length $L = Nb = 286$ mm, viz., $\langle mA \rangle = 3.73$, and the constant mA value is likewise set to $i_0 = mA_0 = 3.73$, such that the scanner-reported $CTDI_{vol}^{TCM}$ is the same for all auto $mA(z)$ distributions and is also the same as $CTDI_{vol}$ at constant mA, such that the scanner-reported value of "$CTDI_{vol}$" (of the first and second kinds) are identical for all four members of this "family." A further constraint (commonality) is imposed on our family: since the number of rotations N and scan lengths $L = \upsilon t_0 = Nb$ are also the same ($N=11$ rotations at intervals $b=26$ mm), they correspond to the *same total* mAs $= \langle mA \rangle t_0 = \langle mA \rangle(N\tau)$, and thus the same *total energy absorbed* (integral dose) E and likewise DLP for all four. The relative mA values (and the resultant doses) are in arbitrary units, and the mA values have been made

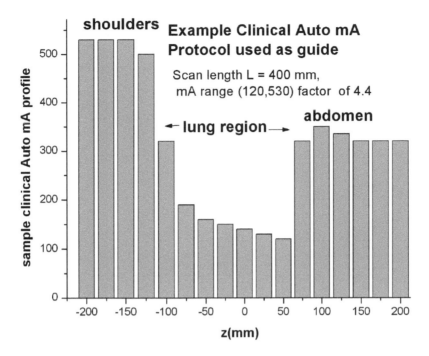

FIGURE 7.3 Realistic example of a clinical auto TCM $mA(z)$ profile, including chest and abdomen in which mA varies by a factor of 4.4 over the scan length.

(From Dixon and Boone, *Medical Physics*, 2013)

to vary in integral steps from 1 to 6 for convenience as noted, but all maintain the same average $\langle mA \rangle = 3.73$ over the scan length of $L = 286$ mm.

By analogy with Figure 7.2, which presents a dose distribution at a constant $mA = 3.73$, Figure 7.4a illustrates the dose distribution for one *shift-variant* member of the auto $mA(z)$ profile family (dubbed *mA3* shown in the figure). The individual $mA(z)$-weighted axial dose profiles $f(z)$ are also shown, as well as the total dose $\tilde{D}_N(z)$ resulting from their summation as described by Eq. (7.9).

This variable $mA(z)$ distribution provides a more compelling example that the local accumulated dose $\tilde{D}_N(z)$ is not proportional to the local $mA(z)$. That is, *the relative $mA(z)$ is seen to drop by a factor of 6*, from a maximum of $mA = 6$ near $z = \pm L/2$, down to $mA = 1$ (over 3 rotations) in the central region, whereas the *relative dose $\tilde{D}_N(z)$ only drops by a factor of* $5/3 \approx 1.7$ from peak value to central ($z = 0$) value (from 5 down to 3). Mathematically, this is due to the convolution in Eq. (7.6a) softening the effect of the mA variation. Physically, it is due to the fact that despite the low local mA at $z = 0$ [$mA(0) = 1$], the scatter tails from the neighboring profiles significantly bolster the central dose at $z = 0$ (even in the face of the steep and sustained mA drop). This is dramatically illustrated in Figure 7.4b in which the same dose data in Figure 7.4a is *re-plotted using a logarithmic scale.* This clearly shows the *vast underpinnings provided by the scatter tails,* as well as their long "reach" over a large fraction of the scan length – every profile contributing a significant dose increment to the dose at $z = 0$.

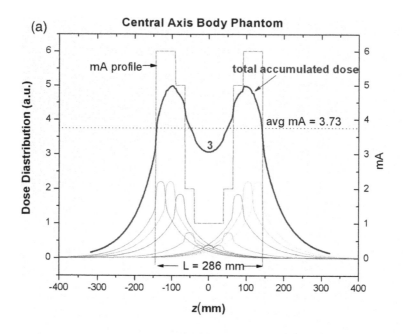

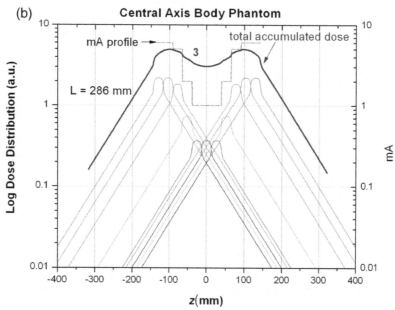

FIGURE 7.4 (a) Accumulated dose obtained from the summation of the $N=11$ individual mA-weighted dose profiles $f(z)$, individually depicted in the figure, based on the $mA(z)$ profile $mA3$ (also plotted), having the common average $\langle mA \rangle = 3.73$ of the entire family of four variable mA profiles. The same scan interval $b = a = 26$ mm used in Figure 7.2 applies here (and in all other examples in Figures 7.5 and 7.6). (b) A logarithmic plot $\log_{10}[\tilde{D}_L(z)]$ of the same data depicted in the linear plot of Figure 7.4a in order to better visualize the "lateral throw" of the scatter tails, which bolster the dose in the center. Thus, despite the fact that the local $mA(z)$ for the 3 central profiles drops by a factor of 6, these scatter tails prop up the central dose and limit its drop at the center to a modest factor of $5/3 = 1.7$ relative to the peak dose.

(From Dixon and Boone, *Medical Physics*, 2013.)

The *mA* profile [thence the *so-called* "$CTDI_{vol}$ per slice" which actually tracks the local $mA(z)$] provides only a qualitative depiction of the accumulated dose profile $\tilde{D}_L(z)$ and can neither predict the magnitude of the maximum dose nor even its *z*-location for TCM. It is recognized that the dose profiles $D_L(z)$ on the peripheral axes will somewhat more closely track $mA(z)$ due to a smaller scatter contribution (smaller scatter-to-primary ratio). In this example, the peripheral axis dose drops by a factor of about 3.8 (vs. 1.7 on the central axis) produced by the *mA* drop of a factor of 6.

The other auto $mA(z)$ models used in this chapter are shown in Figure 7.5 (*mA3* is not re-shown for clarity), each having the same average $\langle mA \rangle = 3.73$ including the constant *mA* profile for which $mA_0 = 3.73$.

The resulting accumulated dose distributions for all four family members are plotted together in Figure 7.6 for comparison – a widely divergent set exhibiting no apparent commonality, *despite the fact that all mA profiles have the same average mA value* $\langle mA \rangle$

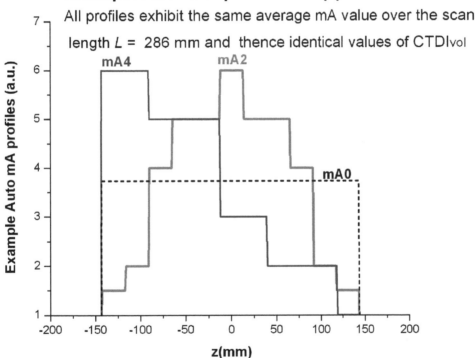

FIGURE 7.5 The other members of the family of *mA* profiles used in this chapter (*mA3* is not re-shown here for clarity), each profile having the same average *mA* value over $L = 286$ mm, namely $\langle mA \rangle = 3.73$, where the constant *mA* value is likewise $mA_0 = 3.73$, such that the scanner will report the same value of "$CTDI_{vol}$" for all *family members* (without making any distinction between $CTDI_{vol}^{TCM}$ for those using TCM with variable $mA(z)$ and the bona-fide $CTDI_{vol}$ of the constant *mA* profile mA_0).

(From Dixon and Boone, *Medical Physics*, 2013.)

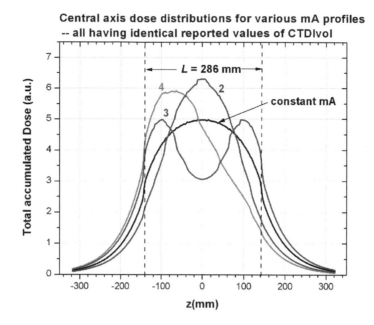

FIGURE 7.6   Accumulated dose distributions for the complete set of 3 auto *mA* distributions from Figure 7.5 – all having the same average *mA* value ⟨mA⟩ = 3.73 taken over the scan length *L*. The actual constant *mA* used is likewise equal to 3.73, hence 4 all have the same $CTDI_{vol}$ ($CTDI_L$ = 5.0) and identical scanner-reported values of "$CTDI_{vol}$." The common number of rotations N = 11 additionally confers the same integral dose *E* and *DLP*.

(From Dixon and Boone, *Medical Physics*, 2013)

and thence the same *scanner-reported* "$CTDI_{vol}$" = 5.0 of the first and second kinds. This value is clearly associated with the constant *mA* accumulated dose distribution (for which $CTDI_L$ = 5.0 predicts the peak dose), but the equal CTDI values have no discernible relationship to the peak dose or distribution of the other TCM distributions. The constant CTDI value might imply, to the unwary, that every dose distribution in Figure 7.6 looks exactly like the symmetric "bell-shaped," constant *mA* profile. But clearly this dose distribution (and $CTDI_{vol}^{TCM}$) has little relevance to the real auto *mA* dose distributions shown, nor can it predict their divergent peak dose values. In short, the scanner-reported $CTDI_{vol}$ has no validity (no apparent connection to) auto *mA* dose distributions (quite unlike an actual constant *mA* which reliably produces the same basic dose distribution for which $CTDI_{vol}$ is always the peak value).

The only commonality conferred *in this example* is that the total area under all the curves in Figure 7.6 (infinite integral) is the same – namely they all represent the same integral dose *E* (same *DLP*), since all have a common scan length *L* = *Nb*. Although all the curves shown in Figure 7.6 have the same total area (infinite integral) we note that *a significant portion of the total energy E deposited* (fraction of the total area) lies *outside the scan interval* (−*L*/2, *L*/2).

7.6.3 Calculation of Average Doses for the Profiles in Figure 7.6 – the Search for Some Commonality of $\mathrm{CTDI}_{vol}^{TCM}$ and $CTDI_{vol}$ for Constant mA

Since $\mathrm{CTDI}_{vol}^{TCM}$ implies nothing about the auto TCM peak dose(s) or their location in Figure 7.6, despite the equality $\mathrm{CTDI}_{vol}^{TCM} = \mathrm{CTDI}_{vol}$ and the fact that all deposit the same energy E (same DLP), perhaps some commonality is implied for the average dose over some region (perhaps over an organ) for the same reported "$CTDI_{vol}$."

The average dose over various 100 mm segments of the scan length exhibited *no evidence of convergence* for the four dose distributions depicted in Figure 7.6. Only when the dose was averaged over the entire scan length does an approximate (albeit anecdotal) convergence occur for our family of four.

Figure 7.7 shows an *approximate convergence* of the average dose over the entire scan length $(-L/2, L/2)$ where $L = 286$ mm, but with large *standard deviations* over this interval as indicated by the error bars (and the *minimum values* far exceed the error bars). *This approximate convergence is not surprising and is clearly related to the absolute constancy of the integral dose E and DLP* (without which no such approximate convergence would occur, and $\mathrm{CTDI}_{vol}^{TCM} = \mathrm{CTDI}_{vol}$ alone would imply nothing). The approximate convergence is to a

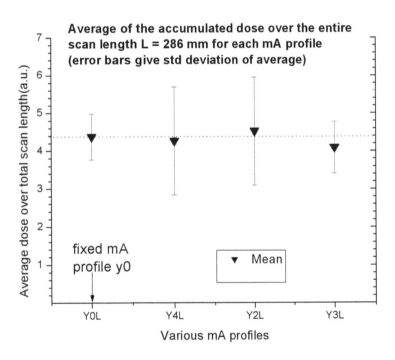

FIGURE 7.7 The average dose over the entire scan length $L = 286$ mm, for the family of mA profiles all having the same average mA, the same $\mathrm{CTDI}_{vol}^{TCM} = \mathrm{CTDI}_{vol} = 5.0$, *and the same scan length* $L = 286$ mm and thence the same integral dose E and DLP [note that the value of $CTDI_{vol}$ (5.0) is above this average as expected]. The end bars indicate the standard deviation.

(From Dixon and Boone, *Medical Physics*, 2013)

dose value of 0.88 $CTDI_{vol}$ *at constant mA* (recall $CTDI_{vol}$ represents the peak dose at $z=0$ and not the average dose over the scan length). A cautionary note: this convergence is based on a $CTDI_L$ and not the $CTDI_{100}$ value imbedded in the scanner-reported $CTDI_{vol}$ which spoils it (Section 7.7).

This convergence is anecdotal for this set of arbitrary *mA* profiles and a long (near equilibrium) scan length, thus we seek a more general confirmation.

However, such an equation for the average dose over the actual scan length (which is clearly lower than the peak dose $CTDI_{vol}$) has never been derived – even at constant *mA*, hence it is instructive to do so in the next section (not to mention pertinent to our present investigation).

7.7 DERIVATION OF SOME NEW ANALYTICAL EQUATIONS TREATING ENERGY AND ACCUMULATED DOSE

The total energy deposited along the central axis z is given by,

$$E = \int_{-\infty}^{\infty} D_L(z)dz = N\int_{-\infty}^{\infty} f(z)dz = LD_{eq} \qquad (7.10)$$

which applies even for sub-equilibrium scan lengths $L=Nb$, and the energy deposited inside the scan length $(-L/2, L/2)$ (the length "directly irradiated" by the primary beam) is given by,

$$E_{in} = \int_{-L/2}^{L/2} D_L(z)dz \qquad (7.11)$$

such that E_{in}/E is the fraction of the energy E absorbed inside the scan interval $(-L/2, L/2)$ which is given by,

$$\frac{E_{in}}{E} = \frac{\int_{-L/2}^{L/2} D_L(z)dz}{LD_{eq}} = \frac{\frac{1}{L}\int_{-L/2}^{L/2} D_L(z)dz}{D_{eq}} = \frac{\overline{D}_L}{D_{eq}} \qquad (7.12)$$

Where $\overline{D}_L$ is the average dose over the scan length $(-L/2, L/2)$.

7.7.1 Restricting the Derivation to the Central Axis and Assuming a Constant Tube Current (*mA*)

The aforementioned integration is facilitated by expressing $D_L(z)$ in a form (due to Li et al. 2012) [equivalent to the convolution format in Eq. (7.1)], which can be expressed in terms of our previously derived Chapter 6 (Dixon and Boone 2010) approach to equilibrium function $H(L)$, namely,

$$\frac{D_L(z)}{D_{eq}} = \frac{1}{2}[H(L+2z)+H(L-2z)] \qquad (7.13)$$

$$H(L) \cong \frac{D_L(0)}{D_{eq}} = \frac{1}{1+\eta} + \frac{\eta}{1+\eta}(1-e^{-L/d}) \tag{7.14}$$

where $d = 117$ mm is the *e-folding* width of the scatter *LSF* and $\eta = 13$ is the scatter-to-primary ratio on the phantom central axis. This gives the result,

$$\frac{E_{in}}{E} = \frac{\overline{D_L}}{D_{eq}} = 1 - \frac{\eta}{1+\eta} \frac{1-e^{-2x}}{2x} \tag{7.15}$$

where $x = L/d$.

This can also be expressed as the average dose over L as a fraction of the peak accumulated dose $D_L(0)$ which is our $CTDI_{\text{vol}}$ surrogate (central axis component, length-corrected from $L = 100$ mm to an arbitrary L) as obtained by using $D_L(0) = H(L)D_{eq}$ as,

$$\frac{\overline{D_L}}{D_L(0)} = \frac{1+\eta\left[1 - \dfrac{1-e^{-2x}}{2x}\right]}{1+\eta(1-e^{-x})} \tag{7.16}$$

Eq. (7.16) represents the ratio of the average dose over the scan length L to the peak dose $D_L(0)$. And finally, the dose at the ends of the scan length $z = \pm L/2$ relative to the peak dose,

$$\frac{D_L(\pm L/2)}{D_L(0)} = \frac{1}{2} \times \frac{1+\eta(1-e^{-2x})}{1+\eta(1-e^{-x})} \tag{7.17}$$

These are tabulated for various scan lengths for *shift-invariant* scan protocols on the central axis in Table 7.1.

Note that the energy deposited *inside* $(-L/2, L/2)$ relative to the *total energy deposited* (E_{in}/E) grows significantly with increasing L.

The ratio of average dose $\overline{D_L}$ to peak dose $D_L(0)$ over $(-L/2, L/2)$ remains remarkably constant with L, with $\overline{D_L}$ remaining about 10% below $D_L(0)$. The value of $\overline{D_L} = 0.88\ D_L(0)$, calculated from Eq. (7.16) for $L = 286$ mm, is equal to that obtained by numerical averaging in Figure 7.7 ($4.4/5.0 = 0.88$), which serves to validate our analytical equation. This is due to the fact that both $D_L(0)$ and $\overline{D_L}$ increase with L at approximately the same rate.

7.7.2 Foiled by $CTDI_{100}$

This constancy could provide a useful physical interpretation of $CTDI_{\text{vol}}^{\text{TCM}}$, but *only if* $D_L(0)$ thence $CTDI_{\text{vol}}$ are computed using the actual scan length L (logical but not the case); unfortunately, $CTDI_w$ and $CTDI_{\text{vol}}$ are based on $CTDI_{100}$, and the scanner-reported value $CTDI_{\text{vol}}$ is based on $D_{100}(0)$ and not $D_L(0)$. Thus $CTDI_{\text{vol}}$ *does not increase with L and it is* the ratio in the last column of Table 7.1 which is applicable, but this truncated $CTDI_{\text{vol}}^{100\,\text{mm}}$ fails us – *causing an apparent 248% variation of the average dose over the scan length over the range of scan lengths shown, and therefore the reported* $CTDI_{\text{vol}}^{100\,\text{mm}}$ *has no utility as a measure of the average dose over the scan length* $\overline{D_L}$.

7.7.3 The IEC Version of a "local $CTDI_{vol}(z)$" and Global $CTDI_{vol}^{TCM}$ for Tube Current Modulation (TCM)

"$CTDI_{vol}(z)$" is not a dose distribution at all, but rather just a scaled $mA(z)$ distribution as given by,

$$CTDI_{vol}^{TCM}(z) = \frac{mA(z)}{\langle mA \rangle_L} CTDI_{vol}^{100}\big|_{\langle mA \rangle} \qquad (7.18)$$

As illustrated previously, the local accumulated dose $D_L(z)$ does not track the local $mA(z)$ – even at constant mA (see Figure 7.2). Indeed, $CTDI_{vol}(z)$ predicts that the "dose" is zero beyond the "directly irradiated" scan interval ($-L/2, L/2$) which Figure 7.2 and Table 7.1 show to be fallacious.

However, the integral of Eq. (7.18) over the entire scan length ($-L/2, L/2$) does yield the *global scanner-reported* $CTDI_{vol}^{TCM} = CTDI_{vol}^{100}\big|_{\langle mA \rangle}$ which is based on the dose over a 100 mm scan length ($CTDI_{100}$) but which is evaluated at the average mA *over the entire scan length L*. The curious physical interpretation of the scanner-reported $CTDI_{vol}^{TCM}$ is the peak central dose for a 100 mm long scan length of an imaginary, constant-current dose distribution having $mA = \langle mA \rangle_L$ which is averaged over the entire scan length.

Li et al. (2014) have utilized our same family of mA distributions; and their resulting calculated dose distributions closely match those shown in Figures 7.4 and 7.6 of this chapter.

Li et al. (2014, 2017) have also used our $L = 286$ mm family of mA distributions illustrated in Figures 7.4 and 7.5 but also extending them to additional cover $L = 100$ mm and 500 mm as well as to the peripheral axes, for *a water phantom diameter of 30 cm*. For $L = 286$ mm, the distributions of Li et al. (2014) on the central axis are quite comparable to ours, as shown in Figures 7.4 and 7.6. For example, the *mA3* distribution shown in Figure 7.4 exhibits a central mA which drops by a factor of 6. Li et al. (2014) show a central drop in dose $D_L(z)$ at $z = 0$ by a factor of 1.8 compared to our data (for the 32 cm PMMA phantom) which shows a similar drop by a factor of 1.7 (both much smaller than the factor of six drop in $mA(z)$). Li et al. (2014) show a drop in dose by a factor of 3.8 on the *peripheral axis* for the same scan – somewhat closer (within 60%) to the factor of six mA drop due to the reduced scatter-to-primary ratio on the peripheral axis. Li et al. (2017) graphically illustrate a wide variation in the ratio of $D_L(z)$ to $mA(z)$ – e.g., for the *mA3* profile, $D_L(z)/mA(z)$ varies by a factor of 5.6/1 on the central axis and 2.3/1 on the peripheral axis; truly supporting the premise that $D_L(z)$ *does not track $mA(z)$* nor the so-called "$CTDI_{vol}(z)$." The *coefficients of variation* (ratio of standard deviation to mean) for the ratio $D_L(z)/mA(z)$ over the $L = 286$ mm scan length for *mA3* were 57% and 27% for the central and peripheral axes, respectively (Li et al. 2017).

But even if these variations were small, that still would not rescue the global scanner-reported $CTDI_{vol}^{TCM}$, (the integral of "$CTDI(z)$" over the scan length), *since it is based on $CTDI_{100}$ which doesn't change with scan length as shown in Section 7.7.*

A change in the basis of $CTDIw$ from $CTDI_{100}$ to $CTDI_L$ would improve the situation on *many fronts*, and is easily affected using available $H(L)$ data [measured (Mori et al. 2005) or using our analytical equations (Dixon and Boone 2010)] as $CTDI_L = [H(L)/H(100)]CTDI_{100}$, in which case the physical interpretation of the reported $CTDI_{vol}^L$ for both auto TCM and constant current would then (and only then) have some commonality and an interpretation, namely: The average dose over the total scan length L is approximately $0.9 \times CTDI_{vol}^L$ for both auto TCM and constant mA (if and only if) $CTDI_{vol}^L$ were calculated using $CTDI_L$ (rather than $CTDI_{100}$). Since most clinical scans exceed $L = 100$ mm, this easily affected scan length correction would also make $CTDI_{vol}^L$ (in and of itself) more closely represent the actual patient dose. Eqs (7.13–7.17) only apply to the central axis of the 32 cm PMMA phantom, however, derivation of like equations for the peripheral axes is straightforward using the equation for $H(L)$ – periphery, previously derived in Chapter 6. The very complete $H(L)$ data by Li et al. referenced in Chapter 5 is also available in parametric equation from a variety of phantom diameters, kVp, etc.

7.8 SUMMARY AND CONCLUSIONS

Rigorous analytical convolution equations describing the accumulated dose distributions $D_L(z)$ for both helical and axial scan trajectories are derived, allowing a side-by-side comparison of $\tilde{D}_L(z)$ for (shift-variant) tube current modulation (TCM) where $i(z) \equiv mA(z)$ [Eq. (7.6a)]; and for $D_L(z)$ for the shift-invariant case using constant tube current (constant mA) [Eq. (7.1) or Eq. (7.6b)], where Eq. (7.1) provides the complete basis for the equations of the CTDI-paradigm. Both equations (7.6a and 7.6b) negate the myth that the local dose $\tilde{D}_L(z)$ or $D_L(z)$ is proportional to the local $mA(z)$ by virtue of their convolution format (and wide scatter tails – as dramatically pictured in Figure 7.4). These convolution equations are complete and their real potential is yet to be tapped. We chose to alternately deploy previously derived (Dixon and Boone 2011) analytic equations for the axial dose profiles in order to provide dose simulations having a strong visual impact, and to bring clarity to the physical underpinnings of the accumulated dose (convolutions will return in Chapter 8).

The convolution in Eq. (7.1) or Eq. (7.6b) representing the CTDI-paradigm produces the familiar symmetric "bell-shaped" dose distribution which has a peak value equal to $D_L(0) = p^{-1}CTDI_L$ (Figure 7.2). Conversely, TCM, via $i(z) = mA(z)$, delivers a wide variety of dose distributions $\tilde{D}_L(z)$ depending on the detailed functional form of $i(z)$ as illustrated in Figure 7.6.

Thus, one can logically conclude the following facts about the scanner-reported values of "$CTDI_{vol}$ of the first and second kinds" (used in IEC 2016); $CTDI_{vol}^{TCM}$ for variable mA and $CTDI_{vol}$ for constant mA.

7.8.1 The Total Energy Absorbed E (Integral Dose) and DLP are Robust between Auto TCM and Constant mA Protocols (but Only for the Same Scan Length $L = \upsilon t_0$)

That is, E and DLP exhibit absolute constancy (are equal) when the reported "$CTDI_{vol}$" values are the same ($CTDI_{vol}^{TCM} = CTDI_{vol}$), whether auto $mA(z)$ or constant mA. Since E and DLP depend only on the total mAs $= \langle mA \rangle t_0$, the detailed functional dependence of $mA(z)$ is not relevant – only its average over L (or t_0) matters for a fixed total "beam-on" time t_0.

7.8.2 The Scanner-Reported Value of "CTDI$_{vol}$" Is Not Robust in the Dose Domain

The dose distribution $\tilde{D}_L(z)$ depends on the exact functional form of $mA(z)$. Thus, unlike the real $CTDI_{vol}$, $CTDI_{vol}^{TCM}$ conveys no information about the maximum dose nor its location along z, but rather $CTDI_{vol} = CTDI_{vol}^{TCM}$ implies a (non-existent) commonality between the dose produced by *an actual constant current* i_0 and an *imaginary* (fantasy) *dose distribution produced by a hypothetical constant current* $i_0 = \langle mA \rangle$. That is, $CTDI_{vol}^{TCM}$ is totally disconnected from the reality of the TCM dose distribution it is supposed to represent. But the fact that $CTDI_{vol}^{TCM}$ lacks relevance is not unexpected – we stated from the outset that CTDI (fundamentally) does not apply to any *shift-variant* protocol such as TCM.

APPENDIX A

The helical dose distribution on a peripheral axis for tube current modulation TCM with $i(z)$:

Dixon has previously shown (Dixon 2003) that Eq. (7.1) and thence Eqs (7.1 and 7.2) – previously derived for a *shift-invariant* helical scan series at constant mA on the central axis – likewise apply to the quasi-periodic helical dose distribution on the peripheral axes if angular smoothing is used.

It is useful to first review this procedure at constant mA shown in Chapter 2 prior to attempting its application to auto mA since it requires a bit of tricky mathematical manipulation. Although the dose rate on a given peripheral z-axis having an angular coordinate θ_z varies with gantry angle θ as a function of $(\theta - \theta_z)$ as a result of the significant variation in primary beam attenuation path length with angle in both the PMMA phantom material and in the bow-tie filter (as depicted in Figure 4.5), where the maximum dose rate occurs when $\theta = \theta_z$, at which gantry angle the attenuation path length in the bow-tie filter and in the phantom itself is a minimum (a penetration depth in the phantom of only 1 cm). However, our previous central axis derivation of Eq. (7.1) based on the constant dose rate on the central axis can still apply to the constant dose rate on a peripheral axis located at θ_z *for a fixed gantry angle* θ, on which the traveling dose rate profile is given as before by $\dot{f}(z - z', \theta - \theta_z) = \tau^{-1} f(z - z', \theta - \theta_z)$ where $z' = \upsilon t$ is the traveling coordinate; and integration of this dose rate profile results in an accumulated dose on that axis (for a fixed gantry angle θ) obtained by analogy with Eq. (7.1) by inspection, namely,

$$D_L(z, \theta_z) = \frac{1}{b} \int_{-L/2}^{L/2} f(z - z', \theta - \theta_z)dz' = \frac{1}{b} f(z, \theta - \theta_z) \otimes \Pi(z/L) \qquad (A.1)$$

Since $z' = \upsilon t$ and $\theta = \omega t$ ($\omega \tau = 2\pi$) are coupled via time as $z' = b\theta/2\pi$, the above integration over z' (or over gantry angle θ) becomes problematic, ruining the chance of expressing it post-integration as a convolution as done in Eq. (7.1) for the central axis dose resulting from the same helical scan, however, by exploiting the dependence of dose rate on $(\theta - \theta_z)$ and its symmetry between θ and θ_z, such that a 2π integration over either is equivalent. Therefore, by taking an angular average of $D_L(z, \theta_z)$ over θ_z in Eq. (A1) at a fixed value of

z, prior to the integration over z' where (unlike θ) θ_z is independent of z'. Re-ordering the integration, this angular average reduces to,

$$D_L(z) = \frac{1}{2\pi} \int\limits_{2\pi} D_L(z, \theta_z) d\theta_z = \frac{1}{b} \int\limits_{-L/2}^{L/2} dz' \frac{1}{2\pi} \int\limits_{2\pi} f(z - z', \theta - \theta_z) d\theta_z$$

$$= \frac{1}{b} \int\limits_{-L/2}^{+L/2} f(z - z') dz' \tag{A.2}$$

which is further based on our recognition of the interior integral over θ_z as simply the axial dose profile on the peripheral axis $f(z)$ evaluated at $z - z'$, with the result that the angular average in Eq. (A2) is identical to Eq. (7.1) previously derived for the phantom central axis. To amplify this somewhat subtle point, the axial dose profile on the peripheral axis results (by definition) from a single axial rotation about a stationary phantom ($\upsilon = 0$ and $z' = \upsilon t = 0$), where $f(z)$ is obtained by averaging the stationary phantom dose rate function $\dot{f}(z, \theta - \theta_z) = \tau^{-1} f(z, \theta - \theta_z)$ by using a 2π integration over either gantry angle θ or over θ_z- the result is the same due to their symmetric appearance in $(\theta - \theta_z)$, and the troublesome coupling between z' and θ in the moving phantom does not apply to the axial dose profile (for which $z' = \upsilon t = 0$). *It is also important to mention that the CTDI-paradigm and the CTDI equation itself can only apply to the peripheral axis dose for a helical scan series by invoking this angular average* to smooth the oscillatory dose (no dose smoothing is required on the central axis where the dose is non-oscillatory and naturally smooth for any pitch). This angular average at a fixed z and the "running mean" (a longitudinal average over a typically small interval $z \pm b/2$ about z used to smooth an axial dose series) have been shown (Dixon 2003) to converge at values of z where dose equilibrium has been established – otherwise they are not quite the same.

Analytical equations have likewise been derived from which to construct the axial dose profile $f(z) = f_p(z) + f_s(z)$ on a peripheral axis.

The peripheral axis helical dose for a TCM technique can be found by applying the same method just used with the same step-by-step procedure using the modified dose rate function now modulated by $mA(z')$, namely $[mA(z')\hat{f}(z - z', \theta - \theta_z)]$ and is now quite straightforward and which is identical to Eq. (7.6a) derived above for the central axis.

This procedure confirms the validity of extending Eq. (7.6a) for $\tilde{D}_L(z)$ to the peripheral axes, (by application of angular smoothing over θ_z), and also validating the use of Eq. (7.7a) for $\tilde{D}_L(0)$ on the peripheral axes.

GLOSSARY

MDCT: multi-detector CT
shift-invariance: translational invariance along z (independent of z coordinate)
AOR: gantry axis of rotation located at isocenter; F = source to isocenter distance

τ: time for single 360° gantry rotation (typically $\tau = 1$ sec or less)

t_0: total "beam-on" time for a complete scan series consisting of N rotations

$N = (t_0/\tau)$: total number of gantry rotations in a scan series (N may not be an integer for helical scanning)

v: table velocity for helical scans

b: table advance per rotation (mm/rot), or *table index*

b: $v\tau$ for helical scans; $b =$ scan interval for axial scans

L: Nb = generalized scan length (axial or helical), $L = v\, t_0 = Nb$ for helical scans

nT: total active detector length referred to isocenter for MDCT (often denoted by "N × T")

a: geometric projection of the z-collimator aperture onto the AOR (by a "point" focal spot); also equal to the *fwhm* of the primary beam dose profile $f_p(z)$. For MDCT $a > nT$ in order to keep the penumbra beyond the active detector length nT (called "over-beaming")

$p = b/nT$: generalized "pitch"

fan beam: a narrow beam of width $a \le 40$ mm

cone beam: typically a beam of width $a > 40$ mm

$\Pi(z/L)$: rect function of unit height and width L spanning interval $(-L/2, L/2)$

$D_L(0)$: accumulated dose accrued at $z = 0$ due to a complete series of N axial or helical rotations covering a scan length $L = Nb$

$f(z)$: single rotation (axial) dose profile acquired with the phantom held stationary consisting of primary and scatter contributions denoted by $f(z) = f_p(z) + f_s(z)$

$f_p(0)$: the dose on the central ray contributed by the primary beam at depth in the phantom

D_{eq}: limiting accumulated dose $D_L(0)$ approached for large $L > L_{eq}$ in conventional CT

$H(L) = D_L(0)/D_{eq}$: approach to equilibrium function

L_{eq}: scan length required for the central dose $D_L(0)$ at $z = 0$ to approach within 2% of D_{eq}

L_{eq}: 470 mm on the central axis of the 32 cm diameter PMMA body phantom

$A_{eq} = (b/a)D_{eq}$: the equilibrium dose constant (independent of aperture a and nT), $A_{eq} = D_{eq}$ for a table increment $b = a$ (no primary beam overlap)

E: the total energy absorbed in the phantom (integral dose)

R: radius of cylindrical phantom

$\hat{a}$: denotes a fan beam aperture (used as necessary to distinguish it from a cone beam aperture a)

η: scatter-to-primary ratio S/P

$i(z) \equiv mA(z)$: x-ray tube current, where $z = vt$

$\langle i \rangle \equiv \langle mA \rangle$: *average mA over total scan length L* as in Eq. (7.4)

$CTDI_{vol}^{TCM}$: scanner-reported value of "$CTDI_{vol}$" for tube current modulation (TCM)

REFERENCES

AAPM Report 204, Size specific dose estimates (SSDE) in pediatric and adult body CT examinations (2011), http://www.aapm.org/pubs/reports/RPT_204.pdf.

Bracewell R.N., *The Fourier Transform and Its Applications*, 3rd ed., McGraw Hill, Boston, (2000).

Dixon R.L., A new look at CT dose measurement: Beyond CTDI. *Med Phys* 30, 1272–1280, (2003).

Dixon R.L., and Ballard A.C., Experimental validation of a versatile system of CT dosimetry using a conventional ion chamber: Beyond $CTDI_{100}$. *Med Phys* 34(8), 3399–3413, (2007).

Dixon R.L., and Boone J.M., Cone beam CT dosimetry: A unified and self-consistent approach including all scan modalities—with or without phantom motion. *Med Phys* 37, 2703–2718, (2010).

Dixon R.L., and Boone J.M., Analytical equations for CT dose profiles derived using a scatter kernel of Monte Carlo parentage with broad applicability to CT dosimetry problems. *Med Phys* 38, 4251–4264, (2011).

Dixon R.L., and Boone J.M., Dose equations for tube current modulation in CT scanning and the interpretation of the associated $CTDI_{vol}$. *Med Phys* 40, 111920, (2013).

Dixon R.L., Munley M.T., and Bayram E., An improved analytical model for CT dose simulation with a new look at the theory of CT dose. *Med Phys* 32, 3712–3728, (2005).

IEC 60601-2-44, Edition 3.2 *Medical Electrical Equipment — Part 2-44: Particular Requirements for the Basic Safety and Essential Performance of X-ray Equipment for Computed Tomography*, International Electrotechnical Commission, Geneva, Switzerland, (2016).

Leitz W., Axelson B., and Szendro G., Computed tomography dose assessment – a practical approach. *Radiat Prot Dosim* 57, 377–380, (1995).

Li X., Zhang D., and Liu B., Equations for CT dose calculations on axial lines based on the principle of symmetry. *Med Phys* 39, 5347, (2012).

Li X., Zhang D., and Liu B., Radiation dose calculations for CT scans with tube current modulation using the approach to equilibrium function. *Med Phys* 41, 111910, (2014).

Li X., Yang K., and Liu B., Characterization of radiation dose from tube current modulated CT examinations with considerations of both patient size and variable tube current. *Med Phys* 44, 5413–5422, (2017).

Mori S., Endo M., Nishizawa K., Tsunoo T., Aoyama T., Fujiwara H., and Murase K., Enlarged longitudinal dose profiles in cone-beam CT and the need for modified dosimetry. *Med Phys* 32, 1061–1069, (2005).

Turner A.C., Zanki, M., DeMarco, J.J., Cagnon, C.H., Zhang, D, Angel, E. . . . McNitt-Gray, M.F. The feasibility of a scanner-independent technique to estimate organ dose using $CTDI_{vol}$ to account for differences between scanners. *Med Phys* 37, 1816–1825, (2010).

Dose Equations for Shift-Variant CT Acquisition Modes Using Variable Pitch, Tube Current, and Aperture, and the Meaning of their Associated $CTDI_{\text{vol}}$

8.1 INTRODUCTION

The CTDI-paradigm is not valid in the case of *shift-variant* techniques for which a parameter varies along z, such as x-ray tube current $i(z)$ modulation (TCM). Other examples include variable table velocity (i.e., variable pitch), and variable z-collimator aperture a, and usage of such *shift-variant* techniques in CT are becoming more common.

TCM is now perhaps more common than constant tube current scan protocols, and variable pitch or pitch modulation (PM) has been implemented in helical shuttles for perfusion studies and in protocols for which pitch is dynamically reduced over certain organs to improve image quality and where concurrent pitch and tube current modulation may come into play.

In Chapter 7, equations were derived representing the accumulated CT dose for tube current modulation (Dixon and Boone 2013) (TCM), finding that the scanner-reported value of $CTDI_{\text{vol}}$ has no physical significance with respect to the TCM dose distribution it is supposed to represent.

The current chapter includes the derivation of similar equations for additional *shift-variant* techniques and parameters, including variable pitch, simultaneous tube current modulation, and pitch variation, and dynamically changing collimator aperture a, as well as scanner-reported dose index values of dubious value due to failure to recognize (or accept) the limitations of *shift-variant* CT acquisition modes. This work includes new convolution simulations for TCM which are validated by comparison with our previous discrete summations (Dixon and Boone 2013).

The primary aim is to present a rigorous theoretical description of the physics from which the dependence on the various scan parameters can be deduced. This forms a basis for a fundamental physical understanding by medical physicists, and which may be useful in aiding manufacturers in the design of *shift-variant* protocols with regard to minimizing patient dose. Additionally, flaws in scanner-reported values of $CTDI_{vol}$ are pointed out and recommended updates to increase their validity (and utility) are included.

8.2 MATERIALS AND METHODS

The traveling axial dose rate profile depicted in Figure 8.1 will be used as a basis for the derivation of the equations as outlined in the next section. The equations for a constant tube current and tube current modulation (TCM) at constant pitch have previously been derived and are outlined for completeness and to set the stage for those that follow. Additionally, the convolution formula for TCM is validated against our previous discrete profile simulations (Dixon and Boone 2013) (they are essentially indistinguishable as illustrated in Figures 8.2 and 8.3). The convolutions are executed using the MATLAB® operator *conv(f,q)*.

8.3 DERIVATIONS

Derivations are performed for the central phantom axis to simplify the development. Extension of all dose equations to the peripheral phantom axes and to concurrent z-axis and angular $i(z, \theta)$ TCM is valid as previously shown in Chapter 7 (Dixon and Boone 2013) and

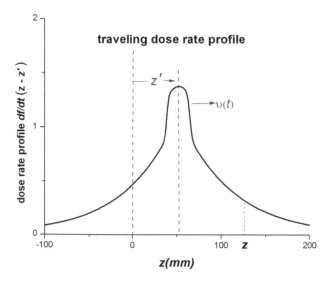

FIGURE 8.1   A traveling *dose rate profile* $\dot{f}(z - z') = \tau^{-1} f(z - z')$ in the phantom reference frame is created when an axial dose profile $f(z)$ is translated along the phantom central axis z by table translation at velocity $\upsilon(t)$, where τ is the gantry rotation period (in sec), and $z'(t) = \int \upsilon(t) dt$.

(From Dixon et al., *Medical Physics*, 2014.)

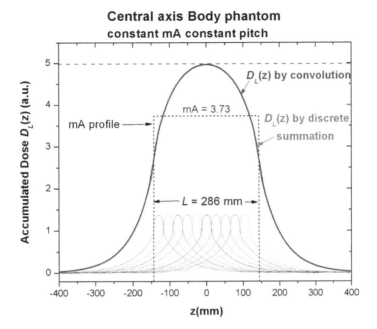

FIGURE 8.2 Accumulated dose at constant *mA* using the convolution Eq. (8.1) for $D_L(z)$ and also using the discrete superposition (summation) of the 11 dose profiles depicted, each having an aperture of $a = 26$ mm and spaced at like intervals using a table increment $b = a = 26$ mm (no primary beam overlap) – the discrete summation distribution being essentially indistinguishable from the convolution. The limiting equilibrium dose $D_{eq} = 5.4$ is first approached at the center $(z = 0)$ for $L \geq 470$ mm and then spreads over a wider range of z as L is further increased (analogous to inflating a balloon against a flat ceiling).

(From Dixon et al., *Medical Physics*, 2014.)

need not be repeated here. A constant projected *z*-collimator aperture width *a* is assumed unless specifically named as a variable. The gantry rotation period τ (in sec) is assumed to be kept constant during the entire scan time t_0 in all cases (variable pitch is assumed accomplished using table speed alone as is the usual case). A glossary of parameters is appended for convenience.

8.3.1 Accumulated Dose for a *Shift-Invariant* Scan (Constant Tube Current and Pitch)

Shift-invariance is a necessary condition for application of the CTDI-paradigm and formula for which no parameters change along *z* during the scan; namely, tube current (*mA*), couch velocity, kV, and *z*-collimator aperture width *a* all remain constant, and the radiation dose-rate profile travels along a phantom *z*-axis at a constant velocity υ by virtue of uniform table translation as depicted in Figure 8.1.

Translation of the table and phantom at constant velocity υ produces a constant *dose rate profile* on the phantom central axis (Dixon et al. 2003) in the form of a traveling wave $\dot{f}(z - \upsilon t) = \tau^{-1} f(z - \upsilon t)$ as depicted in Figure 8.1, where $f(z)$ is the single-rotation (axial) dose profile acquired with the phantom held stationary, and τ is the gantry rotation period

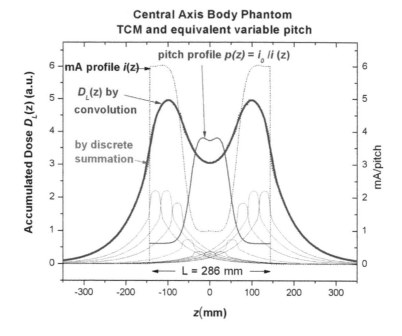

FIGURE 8.3 Accumulated dose obtained using the convolutions: (1) from Eq. (8.5) for TCM for the plotted tube current profile $i(z)$ having $\langle i \rangle = i_0 = 3.73$ [$i(z)$ varies over a relative range of 1 to 6 and has an average value of 3.73] at a constant pitch b_0; or (2) its equivalent for PM in Eq. (8.7) with $b(z) = b_0[i_0/i(z)]$ at constant current i_0 (with the equivalent pitch profile $b(z)/a$ plotted); and (3) a discrete summation of the $N = 11$ mA-weighted axial dose profiles $f(z)$, individually depicted in the figure – their height being based on the same $i(z)$ profile shown and with the same scan interval $b_0 = a = 26$ mm used in Figure 8.2 (and in all other examples).

(From Dixon et al., *Medical Physics*, 2014.)

(in sec). Thus, the dose accumulated at a fixed value of z (also depicted in Figure 8.1) as the profile travels by, is given by the time-integral of $\dot{f}(z - \upsilon t)$ over the total "beam-on" time t_0, namely,

$$D_L(z) = \tau^{-1} \int_{-t_0/2}^{t_0/2} f(z - \upsilon t)dt = \frac{1}{b} \int_{-L/2}^{L/2} f(z - z')dz' = \frac{1}{b}f(z) \otimes \Pi(z/L) \qquad (8.1)$$

the conversion from the temporal to the spatial domain having been made using $z' = \upsilon t$, scan length $L = \upsilon t_0$, and a table advance per rotation $b = \upsilon \tau$, resulting in the convolution equation [Eq. (8.1)] describing the total dose $D_L(z)$ accumulated at any given z-value during the complete scan – the equation of which is the fundamental equation of the CTDI-paradigm from which Eqs (8.2 and 8.3) follow. The accumulated dose at the center of the scan length $(-L/2, L/2)$ is easily obtained by setting $z = 0$ in Eq. (8.1), namely,

$$D_L(0) = \frac{1}{b} \int_{-L/2}^{L/2} f(z')dz' \qquad (8.2)$$

where $CTDI_L$ is equal to $D_L(0)$ for a *table increment* $b = nT$ (pitch $p = b/nT = 1$) and where $CTDI_L = pD_L(0)$.

$$CTDI_L = \frac{1}{nT} \int_{-L/2}^{L/2} f(z')dz' \tag{8.3}$$

These equations have likewise been shown to be valid (Dixon 2003) for axial scanning using the "running mean" (an average over $z \pm b/2$), as well as for helical scans on the peripheral axes using an angular average over 2π at a fixed z.

$D_L(0)$ is the single-axis component of $CTDI_{vol}$ (the latter being arbitrarily truncated to $L = 100$ mm), and nT is superfluous since it cancels out in computing $CTDI_{vol} = p^{-1}CTDIw$ ($p = b/nT$), and only the table increment per rotation b matters.

8.3.2 Accumulated Dose Equation for Tube Current Modulation TCM [Variable $i(z)$, Constant Pitch]

In order to explicitly exhibit the effect of the variable tube current $i(t)$, it can be factored out from the traveling dose rate profile as $\dot{f}(z - \upsilon t, t) = i(t)\tau^{-1}\hat{f}(z - \upsilon t)$, as the product of $i(t)$ with a current-independent shape function $\hat{f}(z)$, namely the axial dose profile per unit current (per unit mA). Since the traveling dose rate function at constant current $\tau^{-1}f(z - \upsilon t)$ implicitly contains a constant current value i_0 as noted, the aforementioned dose rate profile can be modified to the equivalent form,

$$\dot{f}(z - \upsilon t, t) = i(\upsilon t)\tau^{-1}\hat{f}(z - \upsilon t) = \frac{i(z')}{i_0}\tau^{-1}f(z - z') \tag{8.4}$$

and where $f(z - \bar{z}') = i_0\hat{f}(z - z')$ where $z' = \upsilon t$ as before.

The modulated tube current (TCM) accumulated dose $\tilde{D}_L(z)$ *is given by the time-integral of the traveling dose rate profile in Eq. (8.4)*, likewise converted to the spatial domain as before using $z' = \upsilon t$, $L = \upsilon t_0$, and $b = \upsilon\tau$ as shown below,

$$\tilde{D}_L(z) = \frac{1}{b} \int_{-L/2}^{L/2} i(z')\hat{f}(z-z')dz' = \frac{1}{b} \int_{-L/2}^{L/2} \left[\frac{i(z')}{i_0}\right]f(z-z')dz'$$

$$= \frac{1}{b} f(z) \otimes \left[\frac{i(z)}{i_0}\Pi(z/L)\right] \tag{8.5}$$

in which $b = \upsilon\tau$ is the table advance per rotation which is assumed constant, thus b appears outside the convolution integral, but $i(z')$ cannot (pitch $p = b/nT$ is constant).

The local dose $\tilde{D}_L(z)$ at z is not proportional to the local $mA(z) = i(z)$ as is often errone-ously stated (IEC 2016), but rather depends on the current $i(z')$ over all z' locations due to scattered radiation which dominates the dose in CT. That is, the physical interpretation of the product (integrand) $i(z')\hat{f}(z - z')$ is that $i(z')$ determines the peak height of the profile at $z' = \upsilon(t)$, and $i(z')\hat{f}(z - z')$ represents the magnitude of the scatter tail at position z from that z'-profile, the contributions of which must be summed (integrated) over every profile location z'.

Therefore, the accumulated dose in CT at a given location z (including $CTDI_{vol}$ at location $z = 0$) at a given time t depends on *mA past and mA yet to come*, and the scanner-reported "$CTDI_{vol}$ per slice" or $CTDI_{vol}(z)$ does not (and cannot) represent a dose, but merely represents (Dixon and Boone 2013) the relative $i(z) \equiv mA(z)$. In fact, the dose $D_L(z)$ *even at constant mA is not constant* due to variations in such mutual profile scatter. The reader is referred to the many TCM simulations in Chapter 7 to directly visualize the scatter-tail contributions and their effect [particularly to the logarithmic plot (Dixon and Boone 2013) in Chapter 7, Figure 7.4b].

Eq. (8.5) has also been shown in Chapter 7 (Dixon and Boone 2013) to be valid for concurrent angular and z-axis TCM $i(z,\theta)$ using an angular average of *dose* over 2π as well as for both the central and peripheral phantom axes; and likewise reduces to Eq. (8.1) for a constant tube current $i(z') = i_0$ as it must. Note that even at constant current, the variable dose rate on a peripheral axis is not unlike a form of angular TCM – as rigorously treated previously in Chapter 2 using the same angular average over 2π.

8.3.3 Accumulated Dose Equation for Variable Pitch (Pitch Modulation or PM) at Constant Current

In this case, the table velocity υt is varied, and the tube current i_0 and z-collimator aperture a are assumed to be constant in time (thence over z').

Pitch is given by $p = b/nT$ where b is the table increment per rotation, but since nT cancels in every dose equation (including $CTDI_{vol}$), we will treat b as our surrogate pitch (to avoid carrying the superfluous nT), viz., "pitch" varies as $b(t) = \upsilon(t)\tau$ where $\dfrac{dz'}{dt} = \upsilon(t)$ is the variable table velocity and τ is the constant gantry rotation period (in sec).

The traveling dose rate profile on the phantom central axis in Figure 8.1 at time t is given by $\dot{f}(z - z') = \tau^{-1} f(z - z')$, the central location of which at time t is given by $z'(t) = z'(0) + \displaystyle\int_{0}^{t} \upsilon(t)dt$, assuming that the profile passes $z' = 0$ at $t = 0$ and the total beam-on time t_0 covers the symmetric interval $(-t_0/2, t_0/2)$.

The total PM dose $\breve{D}_L(z)$ accumulated at z in Figure 8.1 as the profile completes its transit during the total beam-on time t_0 is obtained by integrating the *dose rate* profile $\dot{f}(z - z') = \tau^{-1} f(z - z')$ over $(-t_0/2, t_0/2)$, the conversion to the spatial domain being made $b(z') = \upsilon(t)\tau$, $dz' = \upsilon(t)dt$ from which $(dt/\tau) = dz'/\upsilon(t)\tau = dz'/b(z')$ and the total scan length is $L = z'(t_0/2) - z'(-t_0/2)$, thus we have for a variable pitch $b(z')$.

$$\breve{D}_L(z) = \tau^{-1} \int_{-t_0/2}^{t_0/2} f(z - z')dt = \int_{z'(-t_0/2)}^{z'(t_0/2)} \frac{f(z - z')dz'}{b(z')} \tag{8.6}$$

8.3.3.1 Table Kinematics for Variable Pitch

The total scan length L is always given by $L = \int\limits_{-t_0/2}^{t_0/2} \upsilon(t)dt = \langle \upsilon \rangle_t\, t_0 = N\langle b \rangle_t\,[z'(t_0/2) - z'(-t_0/2)]$

where $N = (t_0/\tau)$ denotes the total number of rotations, $\langle \upsilon \rangle_t = $ time-averaged table speed, and $\langle b \rangle_t = $ time-averaged table increment (all averaged over the total scan time t_0)

Integrating $\dfrac{dz'}{dt} = \upsilon(t)$ yields $z'(t) = C + \int \upsilon(t)dt$, where C is an arbitrary constant of integration which we can adjust at will. All that matters in the time integration is that the integral covers the time t_0 and we are free to adjust the arbitrary integration constant C (or the origin for $t = 0$) as we wish. Likewise, we can arbitrarily and independently adjust the origin in the spatial integral (if necessary) to the center of the scan length L such that the above spatial integral (which always covers the defined scan length L) also has symmetric integration limits $(-L/2, L/2)$; or equivalently adjust the time origin such that this automatically happens, in which case Eq. (8.6) for variable pitch can always be written in the familiar convolution format,

$$\breve{D}_L(z) = \int\limits_{-L/2}^{L/2} \frac{f(z - z')dz'}{b(z')} = f(z) \otimes \left[\frac{\Pi(z/L)}{b(z)} \right] \qquad (8.7)$$

We would not likely be interested in performing the integration in the time domain as in Eq. (8.6); but rather the convolution format in Eq. (8.7) is both easier to interpret, execute, and is consistent with our other equations for both constant tube current Eq. (8.1) and TCM Eq. (8.5) at a constant table velocity (pitch). It is sufficient to state that an origin shift to force $z' = 0$ in the center of the scan length L is perfectly valid without the details (but the details are readily derivable from the kinematics). The convolution is an *infinite integral* after all.

If we maintain our arbitrary interval of total beam-on time as $(-t_0/2, t_0/2)$ which we previously chose for symmetry purposes, then $C = z'(0)$, namely z' at $t = 0$, such that,

$$z'(t) = z'(0) + \int\limits_{0}^{t} \upsilon(t)dt \qquad (8.8)$$

It is straightforward then to show using Eq. (8.8) that,

$$z'(0) = -\frac{1}{2} \int\limits_{0}^{t_0/2} [\upsilon(t) - \upsilon(-t)]dt \qquad (8.9)$$

Thus, $z'(0) = 0$ if the table velocity $\upsilon(t)$ is an even function of t and thus symmetric about $t = 0$, in which case the spatial integration limits are automatically symmetric about the origin, namely $(-L/2, L/2)$, and we have the simple relationship,

$$z'(t) = \int\limits_{0}^{t} \upsilon(t)dt \qquad (8.10)$$

8.3.3.2 Using a Helical Shuttle as a Specific Example for Illustration of Table Kinematics

Such a symmetric velocity profile exists clinically for the variable pitch protocol called the "helical shuttle mode" for perfusion studies (Siemens and GE) in which the helical pitch is varied between zero at $z = \pm L/2$ to a maximum at the center of the scan length $z = 0$.

The velocity profile $v(t)$ [and pitch $b(t) = v(t)\tau$] for the Siemens helical shuttle can be qualitatively represented as an approximation (for illustrative purposes) by,

$$v(t) = v_m \cos \omega t \qquad (8.11)$$

where $\omega = (\pi/t_0)$, such that,

$$L = \int_{-t_0/2}^{t_0/2} v(t)dt = \langle v \rangle_t t_0 = \frac{2}{\pi} v_m t_0 \qquad (8.12)$$

using Eq. (8.11), and from Eq. (8.12) we have,

$$z'(t) = (L/2)\sin \omega t \qquad (8.13)$$

from which it follows that $b(t) = v(t)\tau$ can be expressed as a function of $z'(t)$,

$$b(z') = v_m \tau \sqrt{1 - \left(\frac{2z'}{L}\right)^2} \qquad (8.14)$$

This represents simple harmonic motion whereby the patient oscillates back and forth through the beam plane (as if on a weak spring). We are only considering the dose delivered in one half-period (over the time t_0) in our equations, since it is identical (and additive) to the dose for subsequent half-periods.

It is important to note that the spatial average of $b(z')$ over $(-L/2, L/2)$ is not equal to the time average of $b(t) = v(t)\tau$ over t_0, viz., $\langle b \rangle_t \equiv \langle b(t) \rangle = \langle v \rangle_t \tau$, where $\langle b \rangle_t$ and $\langle v \rangle_t$ as well as the average pitch $\langle p \rangle_t = \langle b \rangle_t / nT$ *will always refer to a time average over the tube loading time* t_0 and $L = \langle v \rangle_t t_0$ is the total table travel during t_0.

This is readily illustrated using our analytical example $v(t) = v_m \cos \omega t$ where from Eq. (8.12) $\langle v \rangle_t = (2/\pi)v_m$. Since $b(z')$ in Eq. (8.14) represents the top half of an ellipse over $(-L/2, L/2)$ of height $v_m \tau$, it is easy to show using the well-known formula for the area of an ellipse (or by direct integration) that $\langle b(z') \rangle_z = (\pi^2/8)v_m \tau \neq \langle b(t) \rangle_t = (2/\pi)v_m \tau$.

It is straightforward to show in the general case (and confirmed for this specific case) that the significant quantity in the spatial domain (our preferred coordinate system) is the average of $1/b(z)$ [or $1/p(z)$] over the scan length L, namely,

$$\left\langle \frac{1}{b(z')} \right\rangle_z \equiv \langle 1/b \rangle_z = \frac{1}{L} \int_{-L/2}^{L/2} \frac{dz'}{b(z')} = \langle b \rangle_t^{-1} \qquad (8.15)$$

and that scan length is $L = N\langle 1/b \rangle_z^{-1} = N\langle b \rangle_t$, where $\langle b \rangle_t$ denotes the time average by our convention, and where $N = t_0/\tau =$ the total number of rotations (N is not necessarily an integer for helical scans). Also note that $\langle 1/b(z) \rangle_z \neq 1/\langle b(z) \rangle_z$.

Returning to the general case, due to the convolution Eq. (8.7), the local pitch-modulated (PM) dose $\tilde{D}_L(z)$ at point z is not proportional to the local pitch $b(z)$ – a situation quite similar to tube current modulation (TCM) in Eq. (8.5) for which the local dose $\tilde{D}_L(z)$ is not proportional to the local tube current $i(z)$ $[i(z) \equiv mA(z)]$. In fact, comparing the PM dose $\tilde{D}_L(z)$ in Eq. (8.7) for [variable $b(z)$, constant current i_0] to the TCM dose $\tilde{D}_L(z)$ in Eq. (5) [variable $i(z)$, constant pitch $b_0 = \upsilon_0 \tau$] where $f(z)$ in both implicitly contains a constant current $i_0 = \langle i \rangle$, *we see that variable pitch (PM) can mimic variable tube current (TCM) as $i(z) \sim 1/b(z)$*; namely $i(z)b(z) = i_0 b_0$, such *that TCM z-axis tube current modulation can be mimicked by a variation in table velocity (pitch)*. While this duality may not have any practical utility, its theoretical implications are important. If this observation seems obvious – it is not. CT dose is not a local phenomenon due to the dominance of scattered radiation, thence the convolution format of Eq. (8.5) for TCM and Eq. (8.7) for PM in which the accumulated dose at any location z is seen to depend on the tube current $i(z')$ [or the inverse of pitch $b(z')$] *at all locations* z' over the "directly irradiated" scan interval $(-L/2, L/2)$. Indeed, the convolution format for the *shift-invariant* case in Eq. (8.1) describing the CTDI-paradigm produces a bell-shaped dose distribution (Dixon 2003) $D_L(z)$ in which $D_L(z)$ varies by a factor of about two over $(-L/2, L/2)$ despite a constant tube current and constant pitch, and $CTDI_L$ is equal to the peak dose at $z = 0$. This occurs since scatter-tail augmentation (Dixon and Boone 2013) from other dose profiles (rotations) is a maximum at $z = 0$.

Note that the dependence on the inverse of table speed $1/\upsilon(t)\tau = 1/b(z')$ in all derivations arises from the transform from the time to the spatial domain as $dt = dz'/\upsilon(t)$.

8.3.3.3 Summary for TCM and PM

Due to the dominance of scattered radiation in CT dose, for both tube current modulation (TCM) and pitch modulation (PM), the *local dose* at z does not depend solely on the local tube current $i(z)$ or on local pitch $b(z)$, but rather on the tube current $i(z')$, and likewise the pitch $b(z')$ at all locations $(-L/2 \le z' \le L/2)$ via the convolution integral in Eq. (8.5) for TCM or Eq. (8.7) for PM. The other factor $f(z - z')$ in the integrand gives the resulting magnitude of the scatter tail at z when the profile is centered at z' (or both primary and scatter if $|z - z'| \le a/2$ where a denotes the projected z-collimator aperture width and also the primary beam *fwhm*).

8.3.4 Accumulated Dose for Concurrent Tube Current and Pitch Modulation (Concurrent TCM and PM)

This extension may seem intuitively obvious from the aforementioned equations, however, to maintain our rigor, writing the dose rate as in Eq. (8.4) for TCM $\dot{f}(z - \upsilon t, t) = [i(z')/i_0]\tau^{-1} f(z - z')$ but also allowing a variable table velocity such that $z'(t) = C + \int \upsilon(t)dt$, it is straightforward to show that the integral of the dose rate over the total beam-on time t_0 yields,

$$\tilde{D}_L(z) = f(z) \otimes \left[\frac{i(z)}{i_0} \frac{\Pi(z/L)}{b(z)} \right] \qquad (8.16)$$

where $\langle i \rangle = i_0$, and $L = N\langle b \rangle_t = N\langle 1/b \rangle_z^{-1}$, and where i_0 is implicitly imbedded in $f(z)$.

8.4 SUMMARY OF EQUATIONS FOR VARIOUS SCAN PROTOCOLS

The equations in the previous section are summarized in Table 8.1 for convenience.

The CTDI-paradigm is based on Eq. (8.1) in the first row of Table 8.1 (the only *shift-invariant* case) and CTDI has no validity for the other *shift-variant* cases in the table.

8.5 TOTAL ENERGY *E* ABSORBED IN THE PHANTOM (AND *DLP*)

It is straightforward to show from the properties of the convolution below,

$$\int_{-\infty}^{\infty} [f(z) \otimes g(z)]dz = \int_{-\infty}^{\infty} f(z)dz \int_{-\infty}^{\infty} g(z)dz \tag{8.17}$$

that the total energy *E* absorbed in the phantom along (and about) a given *z*-axis *for all cases in* Table 8.1 is given by,

$$E = \int_{-\infty}^{\infty} \tilde{D}_L(z)dz = N \int_{-\infty}^{\infty} f(z)dz = LD_{eq} \tag{8.18}$$

where $N = (t_0/\tau) = $ total number of rotations, such that *E* (and *DLP*) depend (Dixon and Boone 2013) only on the total mAs $= \langle i \rangle t_0 = i_0 t_0$ and the *z*-collimator aperture *a* (*a* = primary beam *fwhm*). That is, *E* and *DLP* are robust (invariant) with respect to *shift-variant* techniques, whereas accumulated dose (including $CTDI_{vol}$) is not (Dixon and Boone 2013).

TABLE 8.1   Summary of Derivation Results These Convolution Dose Equations Are Quite General and Apply to Both the Central and Peripheral Axes of the Phantom as Well as for Concurrent Angular and z-axis Tube Current Modulation i(z,θ) as Previously Shown

Technique	Accumulated Dose Equation	Variable	Table Speed	Table Motion
Constant tube current and pitch Eq. (8.1)	$D_L(z) = \dfrac{1}{b_0} f(z) \otimes \Pi(z/L)$	N/A	υ_0	$b_0 = \upsilon_0 \tau$ $L = \upsilon_0 t_0$
TCM – tube current modulation Eq. (8.5)	$\tilde{D}_L(z) = \dfrac{1}{b_0} f(z) \otimes \left[\dfrac{i(z)}{i_0} \Pi(z/L) \right]$	$i(z')$	υ_0	$b_0 = \upsilon_0 \tau$ $L = \upsilon_0 t_0$
PM – pitch modulation Eq. (8.7)	$\breve{D}_L(z) = f(z) \otimes \left[\dfrac{\Pi(z/L)}{b(z)} \right]$	$b(z')$	$\upsilon(t)$	$b(t) = \upsilon(t)\tau$ $L = \displaystyle\int_{t_0} \upsilon(t)dt = \langle \upsilon \rangle_t t_0$
concurrent TCM/PM Eq. (8.14)	$\tilde{\breve{D}}_L(z) = f(z) \otimes \left[\dfrac{i(z)}{i_0} \dfrac{\Pi(z/L)}{b(z)} \right]$	$i(z'), b(z')$ $\langle i \rangle = i_0$	$\upsilon(t)$	$b(t) = \upsilon(t)\tau$ $L = \displaystyle\int_{t_0} \upsilon(t)dt = \langle \upsilon \rangle_t t_0$
Variable aperture (special case of PM b = a)	$\tilde{\breve{D}}_L(z) = f(z)_a \otimes \left[\dfrac{\Pi(z/L)}{a(z)} \right]$	$b(z') = a(z')$	$\upsilon(t)$	$b(t) = \upsilon(t)\tau$ $a(t) = b(t)$ $L = \displaystyle\int_{t_0} \upsilon(t)dt$

Source: Dixon and Boone, *Medical Physics* (2013) and Dixon et al., *Medical Physics* (2014).

That is, the total energy E deposited (and DLP) are the same, regardless of how the $N = t_0/\tau$ rotations are distributed along z (independent of both L and pitch b) as previously shown in Chapter 7 (Dixon and Boone 2013).

Also note that for TCM at a constant pitch $b_0 = \upsilon_0\tau$, since $z' = \upsilon_0 t$ depends linearly on t, there is no difference between the time and spatial averages of tube current, namely $\langle i \rangle = \langle i \rangle_z = \langle i \rangle_t$, *unlike the two averages for variable pitch*.

8.6 VARIABLE Z-COLLIMATOR APERTURE

It was also be shown in Chapter 6 (Dixon and Boone 2010, 2011) (quite generally) that the total energy absorbed in the phantom (along a given phantom z-axis) in Eq. (8.18) can be expressed as,

$$E = N \int_{-\infty}^{\infty} f(z)dz = Na(1+\eta)f_p(0) \qquad (8.19)$$

where a is the aperture, η is the scatter-to-primary ratio, and $f_p(0) \equiv A_0$ is the primary beam contribution on the central ray to the dose at depth in the phantom; where both η and A_0 are aperture-independent (Dixon and Boone 2011; Dixon et al. 2005) [A_0 depending only on the spectrum (kV, and central ray filtration) and on the mAs per rotation $i_0\tau$]. This result is also independent of the form of $f(z)$. Thus, the total energy E absorbed in the phantom (and DLP) depends directly on the aperture a, and the total mAs ($Ni_0\tau = i_0t_0$) as previously noted. Refining this with the specific form of $f(z) = f_p(z) + f_s(z)$ for the phantom central axis from Chapter 6 (Dixon and Boone 2011),

$$f(z) = f_p(z) + A_0\eta\left[1 - e^{-a/d}\cosh\left(\frac{2z}{d}\right)\right], \quad |z| \le a/2 \qquad (8.20a)$$

$$f(z) = A_0\eta\sinh\left(\frac{a}{d}\right)\exp(-2|z|/d), \quad |z| \ge a/2 \qquad (8.20b)$$

where the terms with $A_o\eta$ represent the scatter contribution $f_s(z)$, and $\exp(-2|z|/d)$ represents the lateral throw of the "scatter tails" ($d = 117$ mm at 120 kVp). [The peripheral axis equations are a bit more complex (Dixon and Boone 2011) but are conceptually the same with respect to integration discussed below]. Integration of the traveling dose rate profile $\dot{f}(z-z') = \tau^{-1}f(z-z')$ pictured in Figure 8.1 using Eqs (8.20a and 8.20b) is complex but possible; however, since the profile height $f(0)_a = A_0\left[1 + \eta\left(1 - e^{-a/d}\right)\right]$ varies non-linearly with a (likewise its shape), hence expression as a spatial integral *in convolution format* is not possible [thus this complex form of *shift-variance* has foiled our previous successes with $i(z)$ and $b(z)$ at constant aperture]. That notwithstanding, variable aperture can be readily handled using the general method described in Chapter 7, Section 7.6 using Eqs (8.20a and 8.20b). However, we note that something interesting occurs when pitch and aperture are simultaneously varied as $a(z') = b(z')$, such that the energy deposited per rotation per unit table increment remains constant; namely, the expected variations of dose

$D_L(z)$ with $a(z')$ or $b(z')$ mutually cancel. If this seems obvious, it is not, since the energy deposited in a single rotation is propagated by scatter and deposited over the entire scan length (and beyond) and is not just deposited locally at the position z in the interval $\Delta z = b(z)$. The analytical proof follows:

Setting $b(z') = a(z')$ in Eq. (8.7) for PM it becomes,

$$\tilde{\tilde{D}}_L(z) = f(z)_a \otimes \left[\frac{\Pi(z/L)}{a(z)} \right] \tag{8.21}$$

Thence,

$$\breve{D}_L(z) - \tilde{\tilde{D}}_L(z) = f(z)_a \otimes \left[\frac{\Pi(z/L)}{b(z)} - \frac{\Pi(z/L)}{a(z)} \right] = 0 \tag{8.22}$$

Q.E.D.

Thus for the helical shuttle, if one were to taper down $a(z')$ in concert with the pitch reduction $b(z')$ as $a(z') = b(z')$, viz., a ratio $a(t)/b(t) = 1$, then one has the equivalent of a *shift-invariant* constant pitch and constant tube current scan, to which the CTDI formula (and thence $CTDI_{vol}$) could be applied (but this would likely complicate the image reconstruction).

Aperture adjustment near the end of helical scan lengths to reduce integral dose ("over-scanning") has little effect on the central dose at $z = 0$ for clinically relevant scan lengths to which $CTDI_{vol}$ refers (at least for *shift-invariant* scans), and using the time-averaged aperture $\langle a \rangle$ would exaggerate its effect on the central dose.

8.7 DOSE SIMULATIONS

All simulations are performed for the central axis of the 32 cm PMMA body phantom at 120 kV. We previously showed graphical simulations for TCM using discrete, axial profile summations and stepped $i(z)$ tube current distributions in Chapter 7 (Dixon and Boone 2013). In order to apply and verify our convolution equations and extend them to variable pitch, we compare the convolutions in Table 8.1 with our previous discrete simulations using the smoothed $i(z)$ distribution applying the same constraints, namely the dose profile $f(z)$ for an aperture $a = 26$ mm from Eqs (8.20a and 8.20b) as depicted in Figure 8.1, with a common total beam-on time t_0 and a common average tube current $\langle i \rangle$; the same total $mAs = \langle i \rangle t_0$, and thence the same total energy E deposited (and same DLP). This means

that the total area under all simulated accumulated dose curves $\int_{-\infty}^{\infty} D_L(z)dz$ is the same.

Additionally, all are constrained to the same scan length $L = \langle v \rangle_t t_0 = N \langle 1/b \rangle_z^{-1} = N \langle b \rangle_t = 286$ mm, where $\langle b \rangle_t$ denotes the time average over t_0; and where $N = t_0/\tau =$ the total number of rotations. We define $p = b/a$ as the relevant *dosimetric pitch* where $a > nT$ for MDCT ("over-beaming"). Thus $\langle b \rangle_t = \langle 1/b \rangle_z^{-1} = 26$ mm, and $\langle p \rangle_t = \langle 1/p \rangle_z^{-1} = 1.0$ in all simulations. The convolutions are executed in MATLAB® using the *conv(f,g)* operator.

8.7.1 *Shift-Invariant* Manual Technique – Constant Tube Current and Pitch

Figure 8.2 depicts the manual technique at constant i_0 and b_0 (constant *mA* and pitch). The continuous convolution and discrete summation [summing the single rotation $f(z)$ profiles depicted, each displaced by $b=a$ from the next by table translation] are essentially indistinguishable – apart from a very small variation at $z=\pm L/2$. Note that constant current and pitch *do not produce a constant dose* – the dose is seen to vary by a factor of two over $(-L/2, L/2)$ on the phantom central axis due to loss of mutual profile scatter near the ends of the scan length.

8.7.2 Variable Tube Current or Variable Pitch

As previously shown, the accumulated dose $\tilde{D}_L(z)$ in Eq. (8.5) for variable $i(z)$ at constant b_0 (pitch) is equivalent to $\tilde{D}_L(z)$ in Eq. (8.7) for variable pitch $p(z)=b(z)/a$ at constant current i_0, if $i(z)b(z)=i_0b_0$, or $b(z)=b_0[i_0/i(z)]$ with $\langle i \rangle = i_0$, and therefore $\langle 1/b(z) \rangle^{-1} = \langle 1/b \rangle_z^{-1} = b_0 = 26\,\text{mm}$. Figure 8.3 illustrates the convolution simulation with $i(z)$ and the equivalent $p(z)=i_0/i(z)$, as well as the previous discrete simulation of dose for $i(z)$ – the discrete summation being essentially equivalent to the convolution, therefore giving us license to use our previous discrete simulations (Dixon and Boone 2013) as examples. This equivalence also confirms the validity of the convolution formulae in Table 8.1.

We note again (Dixon and Boone 2013) that the local accumulated dose $D_L(z)$ does not track the local tube current $i(z)$ [or the local pitch $p(z)$], since the factor of six decrease in $i(z)$ [or an increase in $p(z)$ by the same factor] only produces a drop in the center dose at $z=0$ by a factor of $5/3 = 1.7$.

The discrete summation is again essentially indistinguishable from the convolution – thereby confirming the convolution equations.

In all subsequent simulations, accumulated dose functions correspond to the same *total* $mAs=\langle i \rangle t_0 = i_0 t_0$ where $t_0 = N\tau$ and thence the same total energy E deposited (same area under the curves) and likewise the same *DLP*.

8.7.3 Helical Shuttle (Variable Pitch, Constant Tube Current)

Using our previous qualitative, analytic version of a helical shuttle with table velocity $\upsilon(t) = \upsilon_m \cos \omega t$ and the same constraints on *total mAs* and scan length, the convolution yields the results shown in Figure 8.4 in which the constant pitch dose at the same current i_0 is also plotted for comparison. The total area under both curves is the same (E and *DLP* are the same) and likewise $L = 286$ mm. Note the elliptical shape of $p(z)$ as previously noted [Eq. (8.14)] whereas $p(t) = \upsilon_m \tau \cos \omega t$, so $\langle p \rangle_t = \langle 1/p \rangle_z^{-1} = 1.0$.

8.7.3.1 *Short Helical Shuttle*

Since $L = 286$ mm may be a bit long for a representative perfusion study, we also include an example of a helical shuttle with $\langle p \rangle_t = 1.0$, $L = 138$ mm. Also shown are the alternatives: (1) a constant pitch $p_0 = 1$ helical scan with $L = 138$ mm, or (2) a single rotation of a wide

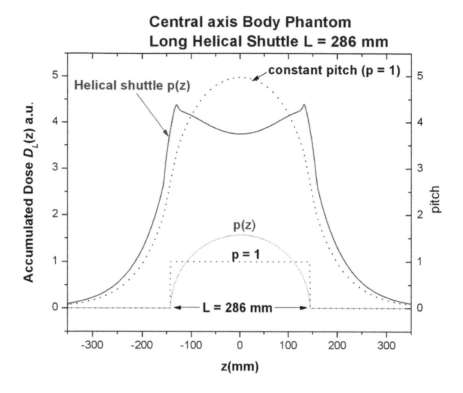

FIGURE 8.4 *Long helical shuttle* at pitch $p(z)$ vs. a helical scan with constant pitch $p = 1.0$ (seen in Figure 8.2), both techniques having $L = 286$ mm and the same reported $CTDI_{vol}$ and DLP, where $CTDI_{vol}$ is derived from the peak central dose $D_L(0) = 5.0$ of the *shift-invariant* (constant current and pitch) technique, and which bears no relation to the helical shuttle dose distribution. Note that the drop-off in dose toward the ends of the scan length at constant pitch due to the loss in scatter is compensated by the boost in dose due to the decrease in pitch, producing a concave distribution compared to the convex "bell-shaped" distribution at constant pitch.

(From Dixon et al., *Medical Physics*, 2014.)

cone beam having $a = 138$ mm about a stationary phantom ($L = 0$) using the theoretical Eqs (8.20a and 8.20b) – the latter two being essentially equivalent as previously proven theoretically and as shown in Chapter 5 (Dixon and Boone 2010) using convolution theory and also as shown in Figure A.1 of the Appendix.

In this case, the decrease in pitch almost exactly compensates for the loss in scatter, producing a nearly flat dose distribution.

By way of validation of the convolution, *the total area under the two dose curves* in Figures 8.4 and 8.5 are confirmed to be the same as required by the convolution (within 0.3% and 0.8%, respectively, by numerical integration).

8.7.3.2 Helical Shuttle vs. Cone Beam

The advantages of the helical shuttle over the stationary table cone beam for perfusion studies is that the non-uniform noise due to the heel effect (and other cone beam artifacts) will be reduced. The noise depends primarily on the transmitted primary beam intensity as

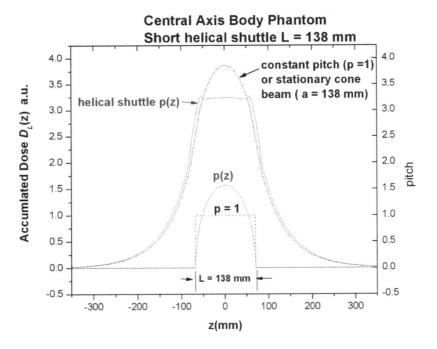

FIGURE 8.5 Short helical shuttle at pitch $p(z)$ vs. helical scan with constant pitch $p = 1.0$, for $L = 138$ mm, both techniques having the same scanner-reported $CTDI_{vol}$ and DLP.

(From Dixon et al., *Medical Physics*, 2014.)

shown in the Appendix (Figure A.1) and not on the dose distribution. For the $a = 138$ mm cone beam, the noise is estimated to vary by 16% over the beam width.

Another dichotomy of the current IEC "$CTDI_{vol}$" reporting system (IEC 2016) is that despite the fact that the dose distribution for the stationary cone beam with $a = 138$ mm is identical to that of the constant pitch helical scan with $L = 138$ mm, the reported values of $CTDI_{vol}$ (Dixon et al. 2014) differ (although both report the same DLP) (see Chapter 9).

8.7.4 Summary of Accumulated Dose Distributions for All Simulated CT Protocols in a Single Figure

The accumulated dose functions plotted in Figure 8.6 all correspond to the same total mAs $= \langle i \rangle t_0 = i_0 t_0$ where $t_0 = N\tau$ and thence the same total energy E deposited (same area under the curves) and likewise the same DLP. Additionally, they are constrained to the same scan length, and all have the same scanner-reported $CTDI_{vol}$ and DLP.

This includes our previous TCM graphical simulations in Chapter 7 (Dixon and Boone 2013) for a family of three different $i(z)$ TCM distributions all having the same $\langle i \rangle = 3.73$ [$i(z)$ varies over a relative range of 1 to 6 with a common average value of 3.73] with $b_0 = a = 26$ mm and $L = Nb_0 = 286$ mm and their corresponding (and identical) variable pitch PM distributions having $b(z) = b_0 \dfrac{i_0}{i(z)} = 26 \, \text{mm} \dfrac{3.73}{i(z)}$, for which $\langle b \rangle_t = b_0 = 26 \, \text{mm}$ and having the same $L = N \langle b \rangle_t = 286 \, \text{mm}$ (not separately plotted since their equivalence to

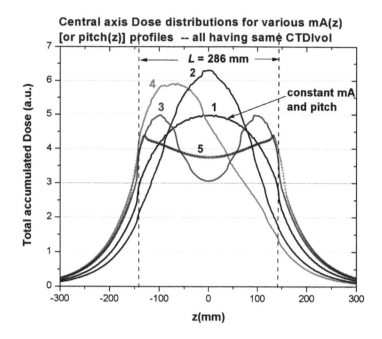

FIGURE 8.6 Accumulated dose distributions for the complete set of TCM $i(z)$ distributions from Chapter 7 (2,3,4) which also double as their corresponding equivalent-pitch distributions having $b(z) = b_0[i_0/i(z)]$. Also shown as (5) is the long helical shuttle; thus Figure 8.6 represents *seven shift-variant* and *one shift-invariant* (1) distributions, all having a common $\langle i \rangle = 3.73$, $\langle b \rangle_t = 26$ mm, scan length $L = \langle \upsilon \rangle_t t_0 = 286$ mm, and the same total mAs $= \langle i \rangle t_0$, thence the same total energy E deposited and same total area under the curves (and likewise the same DLP). *All likewise have identical scanner-reported values of "$CTDI_{vol}$."*

(From Dixon et al., *Medical Physics*, 2014.)

the three TCM distributions was proven previously), and the PM helical shuttle plotted in Figure 8.4. Also shown is the constant current ($i_0 = 3.73$) and constant pitch ($b_0 = 26$ mm) distribution which represents the *lone shift-invariant* case for which the CTDI-paradigm and equation apply.

All eight of these diverse distributions will have the same scanner-reported (IEC 2016) value of $CTDI_{vol}$, only one of which is bona-fide. This is best understood by looking to the source equation of the CTDI-paradigm. The dose distribution $D_L(z) = b_0^{-1} f(z) \otimes \Pi(z/L)$ for a *shift-invariant* technique (constant i_0 and b_0) is the basis of the CTDI-paradigm in Eqs (8.1–8.3) and this convolution for a bona-fide constant tube current i_0 and pitch b_0 always produces a symmetrical "bell-shaped" dose distribution similar to curve (1) in Figure 8.6, which always exhibits its maximum value $D_L(0) = b^{-1} CTDI_L$ at the center $z = 0$, the latter approaching a maximum limiting equilibrium dose D_{eq} at $z = 0$ for scan lengths $L \geq 470$ mm [$D_{eq} = 5.4$ in this case (Dixon and Boone 2013)]. However, Figure 8.6 shows that the *peak doses* for the *shift-variant* TCM or PM distributions (which have the same reported $CTDI_{vol}$) do not necessarily occur at $z = 0$ *and can also exceed $D_{eq} = 5.4$*, which clearly demands a "recalibration" of our physical interpretation of $CTDI_{vol}$.

The IEC methodology for side-stepping a *shift-variant* technique $i(z)$ or $b(z)$ (TCM or PM) is simply replacing ("plugging") the average values $\langle i \rangle$ or $\langle b \rangle_t$ into Eq. (8.3) under the pretext that $\langle i \rangle = i_0$ and $\langle b \rangle_t = b_0$ represent actual (constant) physical values. "Plugging" these "pseudo-constant" values into Eq. (8.1) causes the convolution to generate (without prejudice) the usual "bell-shaped" constant current/pitch dose distribution (albeit an imaginary one) having a peak dose at $z=0$ of $D_L(0) = p^{-1}CTDI_L = 5.0$ (our $CTDI_{vol}$ surrogate), with the result that all eight distributions in Figure 8.6 would simply converge to the "bell-shaped" curve (1), all now having the same peak value $D_L(0) = 5.0$ and identical scanner-reported values of $CTDI_{vol}$. However, this common value of $CTDI_{vol}$ has no discernible connection to the *actual shift-variant* TCM or equivalent PM dose distributions, and therefore little relevance or utility (Dixon and Boone 2013). The IEC methodology has created a "$CTDI_{vol}$ of the second kind."

8.8 SUMMARY AND CONCLUSIONS

Rigorous convolution equations were derived for *shift-variant* tube current modulation (TCM), pitch modulation (PM), and combined TCM/PM techniques as summarized in Table 8.1, as well as for the *shift-invariant* technique (constant current and pitch) – the latter providing the basis for the CTDI-paradigm. It is shown that the variable pitch (PM) $b(z)$ dose distribution at constant current i_0 is equivalent to the TCM $i(z)$ distribution at constant pitch b_0, when $b(z)$ is inversely related to $i(z)$ as $b(z) = b_0[i_0/i(z)]$.

Graphical cumulative dose simulations are provided in Figures 8.2 and 8.3 using both the analytical convolution equations and discrete dose profile summations (Dixon and Boone 2013) – these being essentially congruent, thereby confirming the analytical convolution approach and its physical interpretation which has been more clearly enunciated in this work. Dose distributions for variable pitch (PM) helical shuttles (used clinically for purposes of dose reduction in perfusion studies) are also simulated in Figures 8.4 and 8.5. The *complete set of dose simulations* are combined in Figure 8.6 including all of our previous TCM simulations from Chapter 7 (Dixon and Boone 2013) and their equivalent corresponding (and congruent) PM distributions as well as a helical shuttle – this diverse set (despite having to the same scanner-reported values of $CTDI_{vol}$ and DLP) exhibits no observable commonality in the dose domain.

8.8.1 The Trouble with Scanner-Reported Values of $CTDI_{vol}$ for *Shift-Variant* Scan Techniques

The ad hoc IEC approach (IEC 2016) to handling *shift-variance* by "plugging" the CTDI equation *with the time average of a variable parameter*, averaged over the entire scan time t_0 (or total scan length L); for example $\langle i \rangle, \langle b \rangle, \langle a \rangle, \langle kV \rangle$, results in *scanner-reported values of $CTDI_{vol}$* which have little useful physical significance, nor interpretive utility as previously discussed in detail in Chapter 7 – particularly since they are based on $CTDI_{100}$. However, total energy E and DLP remain robust and transcend *shift-variance*.

Comparing the IEC method to our rigorous equations illustrates that the IEC method is tantamount to replacing $i(z')$ in Eq. (8.5) for TCM or $b(z')$ in Eq. (8.7) for PM by their average values $\langle i \rangle$ or $\langle 1/b \rangle_z^{-1} = \langle b \rangle_t$ and removing them from their respective convolution integrals as constant values. *This is not equivalent to averaging the dose.*

It was previously shown (Dixon and Boone 2013) that the only possible physical interpretation of $CTDI_{vol}$ for a *shift-variant* scan protocol such as TCM or PM with these particular constraints (same E, DLP, and L) is as *an approximation* to the average dose over the scan length. That is, since E/L is the same for all these distributions, one might expect the same average dose over the scan length. That is not the case, however, since a significant fraction of the energy is deposited outside $(-L/2, L/2)$ as seen in Figure 8.6 and in Table 7.1 in Chapter 7. It could, however, serve as a reasonable approximation for longer, clinically relevant scan lengths *but that interpretation is negated* as previously shown in Chapter 7 by the dichotomy that $CTDI_{vol}$ is based on $CTDI_{100}$ whereas $\langle i \rangle$ and $\langle 1/b(z) \rangle$ are averaged over the entire scan length L. This could be easily remedied by a change in the basis of $CTDI_{vol}$ from $CTDI_{100}$ to $CTDI_L$ using the approach to equilibrium function $H(L)$ (Dixon and Boone 2011, 2010). Indeed, at least one manufacturer makes small adaptive-collimation ("tracking") corrections for scan length L to the reported $CTDI_{vol}$ for helical scans (which is inexplicable since $CTDI_{vol}$ represents the dose for $L = 100$ mm), while at the same time ignoring the larger $H(L)$ corrections for increased scatter with scan length. Further, without the aforementioned constraints (same E, DLP, and L), equal values of $CTDI_{vol}$ would imply no useful commonality even with the scan length correction. In short, the total energy absorbed E and DLP remain robust (invariant) for *shift-variant* scan protocols [and for stationary table protocols (Dixon et al. 2014)], but the IEC scanner-reported $CTDI_{vol}$ does not, and its value in these cases should be taken *cum grano salis.*

It is also obvious from the convolution equations in Table 8.1 (and shown by the simulations) that the local TCM dose $\tilde{D}_L(z)$ at point z is not proportional to the local current $i(z)$, and thus averaging the current is not the same as averaging the dose; hence the IEC model of a local "dose per slice" [or "$CTDI_{vol}(z)$"] does not represent *an actual local dose at z* but rather only a relative tube current $i(z) = mA(z)$. In fact, even in the *shift-invariant* case (*constant mA and pitch*), the dose $D_L(z)$ is not constant [as "$CTDI_{vol}(z)$" would predict], but varies by about a factor of two from $z = 0$ to $z = \pm L/2$ on the phantom central axis (Dixon and Boone 2013).

Figure 8.6 also illustrates that the *peak dose* for *shift-variant* TCM or PM techniques is not predicted by the scanner-reported $CTDI_{vol}$; it does not necessarily occur at the center of the scan length; and it may exceed the D_{eq} value of 5.4. This common value of $CTDI_{vol}$ represents the peak, central dose of an imaginary (bell-shaped) dose distribution created under the pretense that the time averages $\langle i \rangle$ and $\langle b \rangle$ behave as *bona-fide* constant current and pitch values.

APPENDIX A

Stationary Cone Beam of *fwhm a* vs. Helical Scan with $L = a$

These are seen to be essentially equivalent as seen in Figure A.1; the small differences are likely due to the effects of anode angle (the heel effect and asymmetric penumbra) which are included in both cone and fan beam profiles but are more pronounced for the wide cone beam having a *fwhm* $a = L$. The noise is seen to vary by about 16% over the slice width due to the heel effect for the wide cone beam.

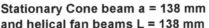

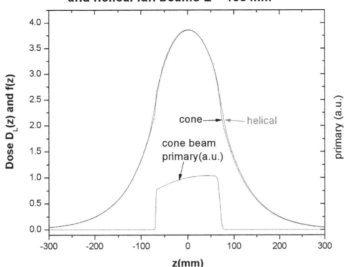

FIGURE A.1 Wide cone beam (Dixon and Boone 2010; Mori et al. 2005) with aperture $a = 138$ mm vs. a helical scan with $L = 138$ mm using an $a = 26$ mm fan beam generated by the convolution in Eq. (8.1). The heel effect for the cone beam primary component is illustrated, showing a variation of the transmitted primary fluence Φ reaching the detectors of 35% and a concomitant noise variation $(1/\sqrt{\Phi})$ of 16% across the slice width (Mori et al. 2005).

(From Dixon et al., *Medical Physics*, 2014.)

GLOSSARY

MDCT: multi-detector CT
shift-invariance: translational invariance of all scan technique parameters along z (independent of z coordinate)
τ: time for single 360° gantry rotation (typically $\tau = 1$ sec or less)
t_0: total "beam-on" time for a complete scan series consisting of N rotations
$N = (t_0/\tau)$: total number of gantry rotations in a scan series (N may not be an integer for helical scanning)
$v(t)$: table velocity for helical scans
$b(t)$: table advance per rotation (mm/rot), or *table index*
$b(t)$: $v(t)\tau$ for helical scans; $b =$ scan interval for axial scans
$\langle y \rangle_t$: time average of any variable $y(t)$ over t_0
$z'(t) = \int v(t)dt$: position of traveling dose rate profile at time t

$$L = \int_{t_0} v(t)dt = \langle v \rangle_t t_0 = N\langle b \rangle_t = N\langle 1/b \rangle_z^{-1} : \text{scan length}$$

$\langle y \rangle_z$: spatial average of the variable $y(z'(t))$ over L
$\langle y \rangle_z \neq \langle y \rangle_t$: in general (tube current being the exception)

$i(z) \equiv mA(z)$: x-ray tube current

$\langle i \rangle = \langle i \rangle_t = \langle i \rangle_z$: average tube current over total scan time t_0 or total scan length L

nT: table advance producing a pitch of unity often denoted by "N×T"

a: "aperture." The geometric projection of the z-collimator aperture onto the AOR (by a "point" focal spot); also equal to the *fwhm* of the primary beam dose profile $f_p(z)$. For MDCT $a > nT$ in order to keep the penumbra beyond the active detector length nT (called "over-beaming")

$p = b/nT$: conventional pitch

$p = b/a$: dosimetric pitch

fan beam: a narrow beam of width $a \leq 40$ mm

cone beam: typically a beam of width $a > 40$ mm

$\Pi(z/L)$: rect function of unit height and width L spanning interval $(-L/2, L/2)$

$D_L(z)$: accumulated dose distribution due to a complete series of N axial or helical rotations covering a scan length L

$f(z)$: single rotation (axial) dose profile acquired with the phantom held stationary consisting of primary and scatter contributions denoted by $f(z) = f_p(z) + f_s(z)$

$f_p(0)$: the dose on the central ray contributed by the primary beam at depth in the phantom

D_{eq}: limiting accumulated dose $D_L(0)$ approached for large $L > L_{eq}$ in conventional CT

$H(L) = D_L(0)/D_{eq}$: approach to equilibrium function – also applies to stationary cone beams as $H(a)$

L_{eq}: scan length required for the central dose $D_L(0)$ at $z = 0$ to approach within 2% of D_{eq}

L_{eq}: 470 mm on the central axis of the 32 cm diameter PMMA body phantom

E: the total energy absorbed in the phantom (integral dose) along and about a given z-axis

η: scatter-to-primary ratio S/P

REFERENCES

AAPM Task Group 111 Report, Comprehensive methodology for the evaluation of radiation dose in X-ray computed tomography, American Association of Physicists in Medicine, College Park, MD, (February 2010), http://www.aapm.org/pubs/reports/RPT_111.pdf.

Dixon R.L., A new look at CT dose measurement: Beyond CTDI. *Med Phys* 30, 1272–1280, (2003).

Dixon R.L., and Ballard A.C., Experimental validation of a versatile system of CT dosimetry using a conventional ion chamber: Beyond $CTDI_{100}$. *Med Phys* 34(8), 3399–3413, (2007).

Dixon R.L., and Boone J.M., Cone beam CT dosimetry: A unified and self-consistent approach including all scan modalities—with or without phantom motion. *Med Phys* 37, 2703–2718, (2010).

Dixon R.L., and Boone J.M., Analytical equations for CT dose profiles derived using a scatter kernel of Monte Carlo parentage with broad applicability to CT dosimetry problems. *Med Phys* 38, 4251–4264, (2011).

Dixon R.L., and Boone J.M., Dose equations for tube current modulation in CT scanning and the interpretation of the associated $CTDI_{vol}$. *Med Phys* 40, 111920, 1–14, (2013).

Dixon R.L., and Boone J.M., Stationary table CT dosimetry and anomalous scanner-reported values of $CTDI_{vol}$. *Med Phys* 41(1), 011907, (2014).

Dixon R.L., Boone J.M., and Kraft R., Dose equations for shift variant CT acquisition modes using variable pitch, tube current, and aperture, and the meaning of their associated $CTDI_{vol}$. *Med Phys* 4, 111906, (2014).

Dixon R.L., Munley M.T., and Bayram E., An improved analytical model for CT dose simulation with a new look at the theory of CT dose. *Med Phys* 32, 3712–3728, (2005).

International Standard IEC 60601-2-44, *Medical Electrical Equipment — Part 2-44: Particular Requirements for the Basic Safety and Essential Performance of X-ray Equipment for Computed Tomography*, 3rd ed., International Electrotechnical Commission, Geneva, Switzerland, (2016).

Mori S., Endo M., Nishizawa K., Tsunoo T., Aoyama T., Fujiwara H., and Murase K., Enlarged longitudinal dose profiles in cone-beam CT and the need for modified dosimetry. *Med Phys* 32, 1061–1069, (2005b).

CHAPTER 9

Stationary Table CT Dosimetry and Anomalous Scanner-Reported Values of $CTDI_{vol}$

9.1 INTRODUCTION

Significant anomalies in the scanner-reported values of $CTDI_{vol}$ for stationary table protocols are described, in which elevated values of $CTDI_{vol}$ over 300% higher than the actual *dose to the phantom* have been observed, and which are well beyond the typical accuracy expected of $CTDI_{vol}$ as a *phantom dose*. Observed clinical anomalies include a neck perfusion study and a high-resolution axial chest CT for which the scanner-reported $CTDI_{vol}$ over-estimated the actual weighted doses by 280% and by 340%, respectively, as calculated herein. Recognition of these types of $CTDI_{vol}$ outliers as incorrect is important to users of CT dose index tracking systems (e.g., ACR DIR); moreover, a solution to the problem is available (Dixon and Boone 2010, 2011; AAPM 2010) and easily implemented as shown in this chapter.

9.2 THE STATIONARY PHANTOM PROBLEM – AND ITS SOLUTION

9.2.1 Analysis Using a Simulation

The simulation illustrating the CTDI-paradigm in Figure 9.1 for a scan length of $L = 286$ mm due to table translation over that distance provides an excellent vehicle for a visual confirmation of the convergence of the CTDI equations to these *non-integral equations* [in the face of a widely held belief that *CTDI* and associated *integral equations* are universally required in CT dosimetry (IEC 2016)]. Figure 9.1 shows a superposition (summation) of 11 identical adjacent dose profiles, corresponding to a z-collimator aperture (primary beam *fwhm*) of $a = 26$ mm, and spaced at like intervals using a table increment $b = a = 26$ mm. The peak accumulated dose at $z = 0$ contributed by the 11 adjacent profiles exhibits a fourfold increase over the peak dose $f(0)$ of a single axial profile due to scatter from adjacent profiles, producing a "bell-shaped" dose distribution $D_L(z)$ with a peak accumulated dose at $z=0$ of $D_L(0)$ related to $CTDI_L$ as $D_L(0) = p^{-1}CTDI_L$. The peak dose of 5.0 obtained from the summation in Figure 9.1 is likewise obtained from the CTDI-paradigm by performing the usual integration of a single profile.

Suppose the phantom of Figure 9.1 had remained stationary (no table motion). Envision $b \to 0$ thence $L = Nb \to 0$, such that all the profiles get closer and closer together and the peak dose grows; and in the limit ($b = 0$, $L = 0$) all $N = 11$ rotations occur at the same fixed location $z = 0$ in the stationary phantom. In this case, all $N = 11$ identical profiles $f(z)$ in Figure 9.1 would simply pile up on top of each other at the location of the profile centered at $z = 0$; producing the *obvious dose distribution* $D_N(z) = Nf(z)$ having a *peak central dose* of $D_N(0) = Nf(0) = 11 \times 1.37 = 15.1$, which is the obvious analog to the peak dose $D_L(0)$ and thence $CTDI_L$ for the moving phantom (as likewise shown by their convergence to same as $b \to 0$, and $L = Nb \to 0$). That is, $f(0)_a$ is clearly the appropriate "dose index" to use for stationary phantom CT dosimetry (Dixon and Boone 2010; AAPM 2010) for narrow fan beams and wide cone beams alike. This solution is apparent by inspection; *no integral equations or pencil chamber are required*; *no dose index is needed*; CTDI does not apply; and $nT \equiv$ "N × T" has no relevance in this case. There is no need to integrate $f(z)$ along the z-axis, since it is already the complete, single-rotation dose distribution function, and $D_N(z) = Nf(z)$ applies to multiple rotations: a study in

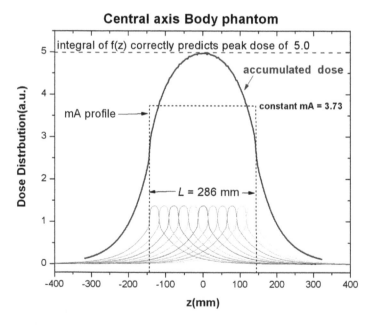

FIGURE 9.1 Accumulated dose for a superposition (summation) of the 11 axial dose profiles depicted, corresponding to aperture $a = 26$ mm and spaced at like intervals using a table increment $b = a = 26$ mm (no primary beam overlap).

(From Dixon and Boone, *Medical Physics*, 2014.)

simplicity compared to the convolution for $D_L(z)$ previously derived in Chapter 2 for "phantom-in-motion" CT dosimetry.

In the case of a stationary table, all that is required (AAPM 2010) is a determination of the peak dose $f(0)_a$ of a single profile for a single aperture setting a via a direct measurement of $f(0)_a$ in a stationary phantom using a 0.6cc Farmer-type thimble chamber for any known aperture setting ($a > 24$ mm) (AAPM 2010). Alternately, a measured (or calculated) value of $CTDI_{100}$ at an aperture a can be corrected as will be described later in Section 9.3.

9.2.2 The Scanner-Reported $CTDI_{vol}$

The above dosimetric simplicity notwithstanding, the International Electrotechnical Commission (IEC 2016) dose standard for N rotations about a stationary CT phantom uses the "phantom-in-motion" formula, namely $CTDI_{vol} = N \times CTDI_w$ (based on $CTDI_{100}$), which is followed by a footnote with the caveat that the above formula will over-estimate the phantom dose due to the inclusion of extra scatter. This is due to the $CTDI_{100}$ formula requiring phantom travel over $L = 100$ mm, whereby the dose profiles are uniformly distributed over $L = 100$ mm at intervals $b = nT$, thus requiring $100/nT$ profiles to fill the 100 mm scan length. For example, in Figure 9.1, $a = 26$ mm, corresponding to $nT = 20$ mm (Mori et al. 2005), thence the $CTDI_{100}$ formula assumes $100/20 = 5$ profiles spaced at intervals of $b = nT = 20$ mm, and the IEC method predicts a dose of $N \times CTDI_{100}$ (based on N rotations of these five profiles about a stationary phantom) rather than using the *real world dose* $Nf(0)$ based on N rotations of the single profile about a fixed $z = 0$. Thus, the IEC formula contains a scatter volume that is 100 mm wide, compared to $a = 26$ mm wide for the actual profile. Moreover, the divisor nT in the $CTDI_{100}$ formula (and thence $CTDI_w$) implies a scan interval (table increment) of $b = nT = 20$ mm which is smaller than the primary beam width ($a = 26$ mm) of these five contiguous profiles; therefore the IEC formula not only over-estimates the scatter (as the footnote warned) but also produces an "over-beaming" dose increase by a factor of $a/nT = 26/20 = 1.3$ (not mentioned in the footnotes); whereas in *the real world of the single, stand-alone profile* $f(z)$ *no such over-beaming exists*. Thus, the use of a $CTDI_{100}$ basis in the IEC formula over-estimates the actual dose on a given phantom axis by a factor of,

$$\frac{CTDI_{100}}{f(0)_a} = \frac{H(100\text{mm})}{H(a)} \frac{a}{nT} \tag{9.1}$$

Where $H(L) = D_L(0)/D_{eq}$ is the approach-to-equilibrium function in Chapters 5 and 6 where D_{eq} is the limiting dose approached for scan lengths $L \geq 470$ mm (Dixon and Boone 2010). The same function also applies to the stationary phantom (Dixon and Boone 2010) as $H(a)$ as shown in Chapter 5 [the formulae for which are given later in Eqs (9.5 and 9.6)]. The factor $H(100)/H(a)$ gives the *over-estimate* due to scatter and a/nT is the over-estimate due to the over-beaming factor.

For the example in Figure 9.1, the scatter increase factor is $H(100)/H(26$ mm$)$ $= 0.61/0.255 = 2.35$, and $a/nT = 1.3$ is the over-beaming factor, thus Eq. (9.1) predicts an over-estimate of the actual *phantom dose* on the central axis $f(0)_a$ by a factor of $2.35 \times 1.3 = 3.1$ (310%); in good agreement (to within 1%) with the ratio of $[CTDI_{100}/f(0)] = 4.3/1.37 = 3.14$ obtained directly from the numerical data of Figure 9.1. Accounting for N rotations, the actual central axis dose is $Nf(0) = 11 \times 1.37) = 15.1$, but the scanner-reported dose is based on a dose for that axis of $N \times CTDI_{100} = 11 \times 4.3 = 47.3$ (310% high as noted). However, one can see from Eq. (9.1) that the scatter-increase factor $H(100)/H(a)$ gets smaller as the beam aperture a approaches 100 mm, and the over-beaming factor may also decrease. The over-estimate on the peripheral axis, likewise calculated from Eq. (9.1) using the peripheral axis $H(a)$ formula (Dixon and Boone 2010) given by Eq. (9.8), is somewhat smaller (168%), thus the weighted (1/3, 2/3) value of $CTDI_{vol}$ reported by the scanner via the IEC formula (IEC 2016)) is still quite high but by a somewhat reduced factor of 2.1 (210%). The CT technical manuals describe many corrections that are applied in the calculation of $CTDI_{100}$ to ensure the accuracy *of the phantom dose* to typically less than $\pm 15\%$ (for acceptance testing). Namely, kV, bow-tie filter, and aperture (over-beaming) corrections are made for both the large and small focal spot, and yet the IEC method produces errors in excess of 300% *in the phantom dose* when the phantom doesn't move as shown in the examples in the next section (when, in fact, a stationary phantom is a much simpler dosimetric problem). Clinical studies using a stationary patient-support table include numerous perfusion studies such as that described in the first example in the following section.

9.2.3 Clinical Examples of Anomalous Values of $CTDI_{vol}$

Clinical examples of anomalous values of $CTDI_{vol}$ which have surfaced in our ACR dose registry are described:

1. *The stationary MIROI (multiple image ROI) phase of a head CTA* using $N = 12$ rotations around the neck with no table motion with $nT = 5$ mm ($a = 7.8$ mm): the scanner-reported value of $CTDI_{vol}$ over-estimated the *actual weighted* dose by 280% (50% of which is due to the inappropriately applied over-beaming factor in the IEC CTDI formula used).

2. *A high-resolution, single rotation axial chest CT* using $nT = 1.25$ mm, $a = 1.84$ mm for which the scanner-reported $CTDI_{vol} = 144$ mGy over-estimated the *weighted dose* by a factor of 3.4 (340%), including an "over-beaming" factor of $a/nT = 1.84/1.25 = 1.47$ (but the over-beaming *dose* penalty does not apply for narrow collimation with a stationary phantom). In this case, the IEC $CTDI_{100}$ formula assumes a *non-existent* "virtual army" of $100/nT = 80$ profiles spaced at $nT = 1.25$ mm with overlapping primary beams (since the formula represents a table advance of $nT = 1.25$ mm whereas the beam width is $a = 1.84$ mm).

Such anomalous values of $CTDI_{vol}$ may also exceed governmental regulatory limits in some countries or states, and require time-consuming explanations. Which begs the question, how many CT users have read the IEC warning footnotes or even have access to IEC publications? Most users of the ACR DIR likely lack the sophistication even to recognize such anomalies – much less to correct them. Given these observations, it is clear that the IEC documents should be updated – particularly the stationary $CTDI_{vol}$ equation.

9.2.4 Total Energy Absorbed E and DLP

The total energy absorbed in the phantom ("integral dose") E and DLP remains robust and unchanged as phantom translation is stopped.

The integral dose for the stationary phantom (Dixon and Boone 2013) $E = N \int_{-\infty}^{\infty} f(z)dz$

is the same as that for the moving phantom (Dixon et al. 2005; Dixon and Boone 2013).

That is, the integral dose (total energy absorbed) due to a single profile is $\int_{-\infty}^{\infty} f(z)dz$ regard-

less of its location along z (or proximity to other profiles), thus E and DLP are unchanged if the scan length $L = Nb$ is increased by increasing the profile spacing b (or pitch); E and DLP will increase only by adding profiles (rotations N) thus increasing the total mAs.

Therefore the same E and DLP formulae are robust and apply for both the stationary and the moving phantom (axial or helical scans), viz., $E = L \times D_{eq}$ and $DLP = L \times D_{100}(0)$ where $L = Nb$, with b canceling in both cases (Dixon and Boone 2013), and,

$$DLP = N \int_{-50}^{50} f(z)dz = (nT)[N \times CTDI_{100}]$$ (in which nT necessarily cancels out) and which

is equivalent to the IEC definition (IEC 2016)) of the stationary table DLP. It is noted that despite the anomalous (elevated) values of scanner-reported $CTDI_{vol}$ for the two aforementioned cases, the reported DLP values remain reasonable and correct. Therefore, for those individuals estimating effective dose from DLP (AAPM 2008) using the "k-factor," this remains a viable approximation.

9.3 APPROPRIATE STATIONARY PHANTOM DOSE EQUATIONS – THE FIX IS EASY

Appropriate stationary phantom analytical formulae (Dixon and Boone 2010, 2011) for $f(z)_a$ and for $f(0)_a$ have been derived in Chapter 6 and it has also been shown that for a wide, stationary cone beam of aperture $a = L$, the stationary phantom and moving phantom dose distributions for scan length L are congruent over all z, i.e., $f(z)_a = D_L(z)$. Using these functions $f(z)_a$, Eqs (9.2–9.5) have been derived for both conventional CT with table translation (on the left) and for stationary phantom dosimetry (on the right) for the body phantom central axis. These equations strip away the integral facade of the CTDI-paradigm, allowing a look into their physical dependencies, and revealing significant commonality. $\hat{a}$ is used for the fan-beam width in conventional CT to distinguish it from a for the stationary phantom.

Conventional CT [nT ≤ 40 mm] L > â   Stationary phantom CT

$$D_L(0) \cong \left(\frac{\hat{a}}{b}\right) f_p(0)\left[1+\eta(1-e^{-L/d})\right]=p^{-1}\text{CTDI}_L \qquad f(0)_a \cong f_p(0)\left[1+\eta(1-e^{-a/d})\right] \qquad (9.2)$$

$$\text{CTDI}_L \cong \left(\frac{\hat{a}}{nT}\right) f_p(0)\left[1+\eta(1-e^{-L/d}))\right] \qquad\qquad \text{Neither CTDI, nor } nT \text{ apply} \qquad (9.3)$$

$$D_{eq} = \left(\frac{\hat{a}}{b}\right) f_p(0)[1+\eta] \qquad\qquad A_{eq} = f_p(0)[1+\eta] = \frac{1}{a}\int\limits_{-\infty}^{\infty} f(z)dz \qquad (9.4)$$

$$H(L) = \frac{D_L(0)}{D_{eq}} \cong \frac{1}{1+\eta}+\frac{\eta}{1+\eta}(1-e^{-L/d}) \qquad H(a) = \frac{f(0)_a}{A_{eq}} \cong \frac{1}{1+\eta}+\frac{\eta}{1+\eta}(1-e^{-a/d})$$

$$\qquad\qquad\qquad\qquad\qquad\qquad\qquad\qquad\qquad\qquad\qquad\qquad\qquad (9.5)$$

---

where $\eta = 13$ is the S/P ratio, and $d = 117$ mm (Dixon and Boone 2010). Peripheral axes equations for $H(L)$ have also been derived (Dixon and Boone 2010) as shown by Eq. (9.8). The major difference is that the peak doses for the moving phantom are all inversely proportional to the table increment per rotation b (or pitch $p = b/nT$), which is non-existent for a stationary phantom. It is also clear that nT simply represents a particular value of table increment $b = nT$. Note that it is the primary beam contribution, $f_p(0)$ common to all Eqs (9.2–9.4), which is the "glue" which holds them together.

The nature of our previously stated correction factor of the IEC formula (IEC 2016)) in Eq. (9.1) is made clear by a left-right comparison of these equations. That is,

$$\frac{H(L)}{H(a)}=\frac{b}{\hat{a}}\frac{D_L(0)}{f(0)_a}=\frac{nT}{\hat{a}}\frac{\text{CTDI}_L}{f(0)_a} \qquad (9.6)$$

Thence, with $\hat{a} = a$

$$\frac{f(0)_a}{\text{CTDI}_{100}}=\frac{nT}{a}\frac{H(a)}{H(100)} \qquad (9.7)$$

Note that nT does not (and cannot) appear in the stationary phantom dose equations on the right [Eqs (9.2–9.4)]. Likewise, nT cancels out in $D_L(0)$ and in $\text{CTDI}_{vol}=p^{-1}\text{CTDI}_w$ for the moving phantom, so its value is moot. Also note that $H(a)=H(L)$ for $L=a$.

On the peripheral axis, $H(a)$ is slightly more complex (Dixon and Boone 2010) ($\eta = 1.5$ and $\varepsilon = 0.305$),

$$H(a) = \frac{1}{1+\eta}+\frac{\eta}{1+\eta}\left[(1-\varepsilon)(1-e^{-a/116})+\varepsilon(1-e^{-a/18.4})\right] \qquad (9.8)$$

The peak dose $f(0)_a$ of a single profile for a single aperture setting a can be determined via a direct measurement of $f(0)_a$ in a stationary phantom using a 0.6cc Farmer-type thimble chamber for any given aperture ($a > 24$ mm) (AAPM 2010; Dixon and Ballard 2007), which can then be scaled to any other aperture a' using the ratio $H(a')/H(a)$, namely,

$$\frac{f(0)_{a'}}{f(0)_a} = \frac{H(a')}{H(a)} \tag{9.9}$$

as indicated by Eq. (9.9).

Alternatively, the absolute stationary peak-dose values $f(0)_a$ can even be determined from the moving-phantom value of $CTDI_{100}$ as illustrated in Eqs (9.2–9.5). Inspection of Eqs (9.2–9.5) shows that obtaining $f_p(0)$ [the primary beam dose component on its central ray at depth in the phantom (ICRU 2012)] is the key to unlocking (and coupling) all these equations – stationary and moving phantom alike, since $f_p(0)$ is common to all the equations and provides a common coupling between them. It is also noted that measurement of $f(z)$ is recommended in ICRU Report no. 87 (ICRU 2012) which contains a rich repository of CT dosimetry data in its Chapter 7.

One can calculate $f(0)_a$ directly for our Figure 9.1 problem (without looking at the figure or its data) using $CTDI_{100}$ in Eq. (9.3) as noted; or more easily using $D_{eq} = 5.4$ in Eq. (9.4). This gives $f_p(0) = 0.386$ and a predicted peak dose $f(0)_a = 1.39$ which agrees to better than 2% with $f(0)_a = 1.37$ determined directly from our profile function. Having $f_p(0)$ allows us to determine anything (Dixon and Boone 2010, 2011) in either the stationary or moving phantom categories.

The ratio of a/nT is available from % geometric dose efficiency (Dixon et al. 2005) $100(a/nT)^{-1}$, either available on the scanner monitor or supplied in the accompanying scanner manual (an IEC requirement). Note that a is also equal to the primary beam full width at half maximum ($fwhm$) usually provided in the scanner documentation. $H(L)$ and $H(a)$ are robust parameters since they represent the ratio of two doses (D_L and D_{eq}) and thus have a very small variation with kV and are independent of scanner make and model (Li et al. 2013).

Moreover, said data for $H(L)$ is readily available in the literature (e.g., Li et al. 2014) for both head and body PMMA phantoms on the central and peripheral axes – both experimental and Monte Carlo simulations. They are available for a series of water and PMMA phantom diameters from 6–55 cm. Li et al. (2013) found that "$H(L)$ has relatively weak dependencies on material (PMMA or water); tube voltage (80–140 kV); and bowtie filter." ICRU report no. 87 (ICRU 2012) also contains a collection of approach-to-equilibrium plots.

9.4 RELATION OF SCANNER-REPORTED DOSE INDICES TO ACTUAL PATIENT DOSE

The CT scanner does not report the actual dose to a given patient. Although the value of the "dose-index" $CTDI_{vol}$ is directly associated with the CT scan performed on a particular patient (say John Smith), it represents the particular type of scan and technique factors used on Mr. Smith. However, its absolute value in mGy is not necessarily representative of the actual dose received by Mr. Smith, even though it may be recorded (with such an implication) in his personal patient dose report. Rather, $CTDI_{vol}$ represents the dose that would be delivered to a 15 cm long plastic disk (phantom) of either 16 cm or 32 cm diameter (head or body) scanned at the same technique used on Mr. Smith, with the exception of the scan length. $CTDI_{vol}$ represents the dose for a scan length of only 100 mm, being calculated from

$CTDI_{100}$. For automatic tube current modulation, $CTDI_{vol}$ is based on the average mA over the entire scan length L whereas $CTDI_{100}$ is based on a 100 mm scan length (a bit of a disconnect). The CTDI-paradigm does not apply for multiple, or single axial rotations about a stationary phantom (such as brain perfusion studies in the cine mode) nor for any *shift-variant* techniques such as the ubiquitous tube current modulation (TCM); hence in these cases the value of $CTDI_{vol}$ reported by the scanner is not representative of the dose – even to a *phantom*.

For a body scan, the actual dose to a thin patient will be much larger than that for a thick patient for the same manual scan technique (kVp, mAs, pitch, etc.), whereas the reported value of $CTDI_{vol}$ is exactly the same for both. Thus, the common value of $CTDI_{vol}$ reported by the scanner in mGy is not likely to represent the dose to either, but rather represents the dose to their dosimetry surrogate. Namely, a 32 cm diameter plastic body phantom which is supposed to represent the body habitus of *every patient who gets a body scan*, whether thick or thin or whether receiving an abdomen or lung scan.

The primary use of $CTDI_{vol}$ is therefore not as an absolute patient dose to the patient being scanned, but rather as a relative dose indicator – to assist the CT operator in evaluating the *relative dose implications* of various choices of CT scan parameters available, and thus to avoid the often unnecessary use of high dose techniques. It is not a measure of "machine output" as is sometimes stated – it depends on pitch which has nothing to do with "machine output."

Although the reported dose $CTDI_{vol}$ is by inference directly associated with an individual patient, it is a very crude measure of the actual dose to that patient, so its absolute value is of secondary importance in that regard. However, the value of $CTDI_{vol}$, together with other patient-specific information, may be quite useful to the medical physicist in reconstructing a more accurate (*albeit still approximate*) patient dose when such a dose reconstruction is specifically requested. An example would be computing a fetal dose for a pregnant patient receiving a CT scan.

9.4.1 Size-Specific Dose Estimates (SSDE)

The basic SSDE dose index concept presented in the Reports of AAPM Task Group 204 and 220 (AAPM 2011, 2014) is an approach to develop a more reasonable *estimate* of patient dose using the scanner-reported $CTDI_{vol}$ and conversion factors that account for patient size. In situations where a fixed tube current is employed, and the patient anatomy and circumference are reasonably homogeneous over an entire CT scan, the basic SSDE provides an improved *estimate* of dose as compared to $CTDI_{vol}$. If the *average* effective diameter d is known for a given patient over the anatomy scanned, the SSDE is determined as,

$$SSDE = f_d CTDI_{vol} \qquad (9.10)$$

where f_d converts $CTDI_{vol}$ in air kerma to the absorbed dose to water, viz., to SSDE. Thus, a small patient will correctly be attributed a relatively higher radiation dose compared to $CTDI_{vol}$ due to reduced attenuation compared to the CTDI phantom. Likewise, the SSDE associated with an exam of a *very large* patient may be relatively lower than the $CTDI_{vol}$ value. The SSDE concept also takes some consideration of the scan length into account via the use of typical scan lengths for clinical exams. The effective patient diameter is determined from the pre-scan scout projection views. The conversion factors f_d for all body scan protocols are based on $CTDI_{vol}$ for the 32 cm phantom – even for pediatric body scans; however, caution must be exercised since some older scanners may report $CTDI_{vol}$ for pediatric body scans based on the 16 cm diameter phantom (an older IEC standard) leading to an untoward SSDE about 200% high.

The geometric estimation of effective patient diameter from the pre-scan scout projections has limitations, e.g., it does not account for decreased attenuation in lung tissue when present. As a consequence, AAPM Task Group 220 Report (AAPM 2014) has suggested a more complex attenuation-based method for SSDE based on axial scan reconstruction data. The American College of Radiology Dose Index reporting system (ACR DIR) not only captures $CTDI_{vol}$ and DLP data but also computes and reports SSDE.

The IEC is currently working on developing a *scanner-reported* value of SSDE, *based on AAPM 2014*, which may be coming soon to a scanner near you.

9.4.2 Anomalous Values of SSDE

A faulty value of $CTDI_{vol}$ such as those for stationary table techniques and *shift-variant* scan techniques such as tube current modulation (TCM) as previously discussed will likewise result in a faulty SSDE. In these cases, DLP will be the more robust dose index. For example, our dose –index tracking software (ACR DIR) – showed our median $CTDI_{vol}$ and thence SSDE value for a routine chest exam to be well above the national average. This was discovered to be due to the inclusion of three well-separated, high-resolution, narrow beam axial scans to the helical scan protocol – each of these having been previously shown to give a 340% over-estimate of the weighted average dose via $CTDI_{vol}$. However, DLP was in normal limits as illustrated in Figures 9.2a and b.

Unfortunately, the fact that SSDE is based on typical clinical body scan lengths of $L > 100$ mm, the SSDE conversion adds even more extra (non-existent) scatter to these *narrow beam, stationary table* techniques – further increasing this over-estimate by a factor of $H(L)/H(100)$.

Hopefully the previous discussion of stationary-table and *shift-variant* techniques will alert the user to these faulty $CTDI_{vol}$ and SSDE values.

9.5 $CTDI_{100}$ FOR WIDE BEAMS (IEC VERSION) – CRACKING THE "CTDI ENIGMA" CODE

The paper "The Trouble with $CTDI_{100}$" (Boone 2007) showed a significant drop in the value of $CTDI_{100}$ as primary beam widths grew comparable to (and even larger than) the length of the 100 mm pencil chamber. This spurred the IEC to issue an "empirical patch" to the $CTDI_{100}$ for $nT > 40$ mm based on measurements of $CTDI_{\text{free-in-air}}$ made using the 100 mm long pencil chamber, which can be simplified as,

$$CTDI_{100} = \frac{(nT/a)_{\text{ref}}}{(nT/a)}(\text{CTDI}_{100})_{\text{ref}} \qquad (9.11)$$

Where the reference value for $CTDI_{100}$ is taken at $nT = 40$ mm or less. This "patch" was designed to keep $CTDI_{100}$ equal to the same fraction of the equilibrium value of $CTDI_{100}$ for larger beam widths, namely,

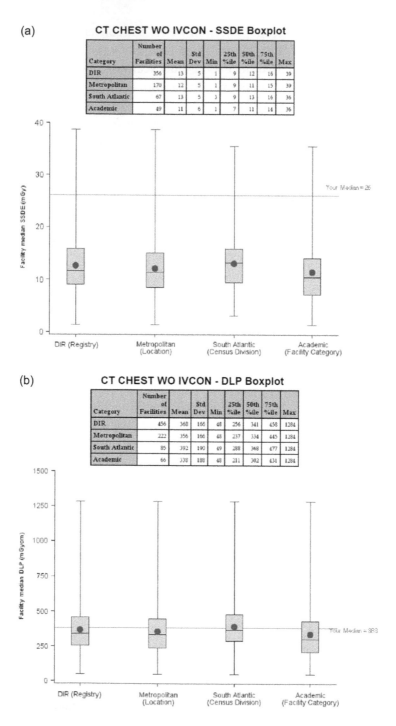

FIGURE 9.2 (a) Anomalous SSDE based on a faulty $CTDI_{vol}$ from the American College of Radiology Dose Index tracking registry. The horizontal line indicates the median for our facility. (b) The median DLP for the same protocol as that shown in Figure 9.2a remains robust and indicates no problem.

(From Dixon, 2018.)

$$CTDI_{100} = H(100)CTDI_{\infty} \tag{9.12}$$

where $H(100)$ is the approach-to-equilibrium function.

When first presented at an IEC MT30 CT committee meeting, it puzzled most of us, hence I dubbed it "CTDI enigma," but soon realized that since $CTDI_{\text{free-in-air}}$ tracks the aperture a (actually a/nT) that this formula was likely related to the constancy of CTDI-aperture ($CTDI_a$) previously introduced in Chapter 4 (Dixon et al. 2005). So here it is in a nutshell – just take some ratios based on the following:

$$CTDI_a = \frac{1}{a}\int_{-\infty}^{\infty} f(z)dz = \frac{nT}{a}CTDI_{\infty} = \text{const.} \tag{9.13}$$

$$CTDI_{100} = H(100)CTDI_{\infty} \tag{9.14}$$

$$CTDI_{\text{free-in-air}} = f_p(0)\frac{a}{nT} \tag{9.15}$$

Thus the "corrected" $CTDI_{100}$ is,

$$\frac{CTDI_{100}}{(CTDI_{100})_{\text{ref}}} = \frac{(nT/a)_{\text{ref}}}{(nT/a)} \tag{9.16}$$

So, you don't need to make any free-in-air pencil chamber measurements to obtain this ratio. The aperture a (primary beam $fwhm$) is readily available from the scanner technical manual ("accompanying documents") which may include an "aperture correction" (a/nT) to $CTDI_{100}$ or a dose efficiency (nT/a) value for the scan. Not only that, but the free-in-air measurement itself becomes problematic when the beam width exceeds 100 mm, requiring one to make two pencil chamber measurements "end-to-end," as it were (or use a longer pencil chamber).

This gives an improved "table-in-motion" value for $CTDI_{100}$ at wide beam widths; however, larger beam widths are more often related to stationary table cone beam CT (SCBCT) protocols to which the CTDI-paradigm does not apply. We have seen its failure previously in this chapter for narrow beams, so how well does it work for wide beams incident in a stationary phantom?

Not particularly well, as shown in Table 9.1 using the measured data of Mori et al. 2005 and the $nT = 32$ mm beam as the reference. The modified $CTDI_{100}$ does maintain a constant ratio of 0.61 with $CTDI_{\infty}$ as it was designed to do; however, it fails to predict the stationary phantom central dose $f(0)$ across the gamut of beam widths shown.

Contrast that with our suggested method of correcting $CTDI_{100}$ to predict the stationary phantom dose $f(0)_a$ given by Eq. (9.7) and shown in Table 9.2. The resulting value of the calculated $f(0)_a$ is within ±2% of the measured value. The primary beam contribution $f_p(0)$, [computed using the measured $f(0)_a$ values and by setting the S/P ratio $\eta = 0$ in Eq. (9.2)] is shown in Table 9.2 to be constant (within ±2%) as one would expect. The primary component is a small fraction of the total peak height $f(0)_a$ – the fraction growing smaller

TABLE 9.1   Measured Data (Mori et al. 2005) – Central Axis, Body Phantom Compared to the IEC Value Corrected for Beam Width using Eq. (9.11)

Aperture a(mm)	nT(mm)	a/nT	CTDI-aperture $CTDI_a = \dfrac{1}{a}\displaystyle\int_{-\infty}^{\infty} f(z)dz$	$CTDI_\infty$	$CTDI_{100}$	$CTDI_{100}c$ Corrected	Measured Central-ray Dose $f(0)a$	Ratio $\dfrac{CTDI_{100c}}{CTDI_\infty}$
138	128	1.08	6.14 mGy	6.63 mGy	4.04 mGy	4.07 mGy	4.37 mGy	0.61
111	96	1.16	6.18	7.17	4.37	4.37	3.90	0.61
80	64	1.25	6.22	7.78	4.75	4.71	3.19	0.61
49	32	1.53	6.18	9.46	5.77	5.77	2.27	0.61

TABLE 9.2   Measured Data (Mori et al., 2005) – Central Axis, Body Phantom Compared to the Valid Correction Factor in Eq. (9.7)

Aperture a(mm)	nT (mm)	a/nT	$\dfrac{H(a)}{H(100)}$	$\dfrac{H(a)}{H(100)}\dfrac{nT}{a}$	$CTDI_{100}$	Computed Central-ray Dose $f(0)a$	Measured Central-ray Dose $f(0)a$	Primary Beam Contribution $fp(0)$
138	128	1.08	1.18	1.093	4.04 mGy	4.42 mGy	4.37 mGy	.442 mGy
111	96	1.16	1.06	0.914	4.37	3.99	3.90	.443 mGy
80	64	1.25	0.843	0.674	4.75	3.20	3.19	.429 mGy
49	32	1.53	0.595	0.389	5.77	2.24	2.27	.412 mGy

with an increasing beam width a as the scatter component builds. We again note that $f_p(0)$ is the common connector in all Eqs (9.2–9.4) relating table-in-motion to stationary table dosimetry.

The computed central-ray dose $f(0)_a$ in Table 9.2 is obtained by correcting $CTDI_{100}$ using Eq. (9.7). The primary beam contribution is computed from Eq. (9.2) by setting $\eta = 0$ and using the measured value of $f(0)_a$.

9.6 SUMMARY AND CONCLUSIONS

The integral equations of the CTDI-paradigm (including $CTDI_{vol}$) cannot apply to stationary phantom/table dosimetry, and have been shown to apply only to a moving phantom (axial or helical scans with table translation) – *the motion of which is the fundamental source of the integral format*. The use of the $CTDI_{100}$ equation given in the IEC 2016 standards for the case of a stationary table leads to elevated scanner-reported values of $CTDI_{vol}$ which have been shown to exceed the actual weighted dose to the phantom by 300% or more. Recognition of these outliers as "bad data" is important to users of CT dose index tracking systems (e.g., ACR DIR), and *a method for recognition and correction of same is provided by Eq. (9.7)*. The abnormally high dose values discussed herein are more likely to occur with the use of narrow fan beams ($nT \leq 40$ mm) for *stationary table procedures* such as perfusion studies or narrow beam axial chest cuts, as illustrated in our clinical examples. It was also shown that the IEC correction for $CTDI_{100}$ for wide beams [Eq. (9.1)] does not apply in the case of stationary cone beam CT and does not predict the central peak dose as shown in Table 9.1.

Therefore, one should regard the scanner-reported $CTDI_{vol}$ for any stationary table procedure with suspicion and rather rely more on DLP (which remains robust). This anomaly will likely be more prevalent in stationary table exams using narrow fan beams ($nT \leq 40$ mm), with the magnitude of the dose over-estimate increasing with decreasing nT (thence decreasing aperture a) as illustrated by Eq. (9.7) and by our clinical examples.

The proper equations for stationary phantom dosimetry $D_N(z) = Nf(z)$ where $N =$ number of rotations and the "peak dose" $Nf(0)$ have been described herein; these being much simpler *with no integral equations or pencil chamber required; no dose index needed; CTDI does not apply; and $nT \equiv$ "N × T" has no relevance.*

REFERENCES

AAPM Report No. 96, The measurement, reporting, and management of radiation dose in CT, American Association of Physicists in Medicine, College Park, MD, (January 2008).

AAPM Report No. 204, Size-specific dose estimates (SSDE) in pediatric and adult body CT examinations, American Association of Physicists in Medicine, College Park, MD, (2011).

AAPM Report No. 220, Use of water equivalent diameter for calculating patient size and size specific dose estimates (SSDE) in CT, American Association of Physicists in Medicine, College Park, MD, (2014).

AAPM Task Group 111 Report, Comprehensive methodology for the evaluation of radiation dose in x-ray computed tomography, American Association of Physicists in Medicine, College Park, MD, (February, 2010), http://www.aapm.org/pubs/reports/RPT_111.pdf.

Boone J.M., The trouble with $CTDI_{100}$. Med Phys 34(4), (2007).

Dixon R.L., Radiation dose in computed tomography in Handbook of X-ray Imaging, ed. P. Russo, 791–804, CRC Press, Taylor and Francis, (2018).

Dixon R.L., and Ballard A.C., Experimental validation of a versatile system of CT dosimetry using a conventional ion chamber: Beyond $CTDI_{100}$. Med Phys 34(8), 3399–3413, (2007).

Dixon R.L., and Boone J.M., Cone beam CT dosimetry: A unified and self-consistent approach including all scan modalities—with or without phantom motion. Med Phys 37, 2703–2718, (2010).

Dixon R.L., and Boone J.M., Analytical equations for CT dose profiles derived using a scatter kernel of Monte Carlo parentage with broad applicability to CT dosimetry problems. Med Phys 38, 4251–4264, (2011).

Dixon R.L., and Boone J.M., Dose equations for tube current modulation in CT scanning and the interpretation of the associated $CTDI_{vol}$. Med Phys 40, 111920, 1–14, (2013).

Dixon R.L., Boone J.M., Stationary table CT dosimetry and anomalous scanner-reported values of $CTDI_{vol}$. Med Phys 41, 011907, (2014).

Dixon R.L., Munley M.T., and Bayram E., An improved analytical model for CT dose simulation with a new look at the theory of CT dose. Med Phys 32, 3712–3728, (2005).

ICRU, Report No. 87 of the International Commission on Radiation Units and Measurements, Patient dose and image quality in computed tomography, Oxford University Press, London, (2012).

International Standard IEC 60601-2-44, Medical Electrical Equipment — Part 2-44: Particular Requirements for the Basic Safety and Essential Performance of X-ray Equipment for Computed Tomography, 3rd ed., International Electrotechnical Commission, Geneva, Switzerland, (2016).

Li X., Zhang D., and Liu B., Monte Carlo assessment of CT dose equilibration in PMMA and water cylinders with diameters from 6 to 55 cm. Med Phys 34, 3399–3413, (2013).

Li X., Zhang D., and Liu B., Longitudinal dose distribution and energy absorption in PMMA and water cylinders undergoing CT scans. Med Phys 41, 10, (2014).

Mori S., Endo M., Nishizawa K., Tsunoo T., Aoyama T., Fujiwara H., and Murase K., Enlarged longitudinal dose profiles in cone-beam CT and the need for modified dosimetry. Med Phys 32, 1061–1069, (2005).

CHAPTER 10

Future Directions of CT Dosimetry and A Book Summary

10.1 BEYOND CTDI

10.1.1 Estimation of Organ Doses

There is a growing movement to calculate individual organ doses in CT, primarily based on Monte Carlo simulations, which begs the question: What are we to do with such data? Even if we could calculate organ doses accurately, are the risk factors for the individual organs that well known? Or will they even be?

10.1.1.1 Tube Current Modulation (TCM) and SSDE

Presently, the most common mode of performing CT examinations is helical ("spiral") scanning employing TCM. $CTDI_{vol}$ is determined from the average tube current (mA) used during the entire scan, and as a consequence the basic SSDE will yield an estimate of patient dose as if a scan had been performed with a fixed mA equal to the average mA. Thus, any local variations in patient exposure from using TCM will not be translated by the basic SSDE, and the oft-made assumption that the local dose is proportional to the local $mA(z)$ using a "$CTDI_{vol}(z)$" and an "$SSDE(z)$" is flawed as previously discussed in Chapter 9 (scatter is appropriately accounted for by using the tube current in *a convolution integral* with the dose profile as shown in Chapter 5). Indeed, "$CTDI_{vol}(z)$" predicts zero dose outside the scan interval ($-L/2$, $L/2$) whereas a significant fraction of the energy is deposited outside this interval by scatter as previously illustrated in Table 7.1; e.g., for a scan length of 100 mm, 44% of the energy is deposited outside of the directly irradiated length $L = 100$ mm on the central phantom axis (Table 7.1).

A recent paper (Tian et al. 2016) employing such a convolution method reported an improved organ dose accuracy over the method which assumes a local dose proportional to $mA(z)$.

Once organ doses have been calculated, then what? Papers featuring organ dose computations rarely (if ever) apply the currently accepted organ risk factors to compute overall risk, although (as a reviewer) I have suggested that they do so. These risk factors are age (and sex) related. How is organ dose information of value – either to the patient or the physician?

Some *commercial dose-tracking software* now include an organ-dose computation for each patient; for example, by matching the patient's body habitus to a particular humanoid phantom on which Monte Carlo calculations of organ dose have been made. If these are further normalized to the patient, based on the scanner-reported value of $CTDI_{vol}$, then the above-mentioned caveats concerning $CTDI_{vol}$ remain in play.

The IEC is currently working on a model by means of which SSDE will additionally be reported by the scanner – based on a water-equivalent patient diameter d (AAPM 2011, 2014), and once again using $CTDI_{vol}$ as a basis, and which may soon be coming to a CT scanner near you. The various CT manufacturers will be responsible for the methodology (and validation of) the computation of water-equivalent diameter d, and thence SSDE.

10.1.2 Understanding Risks from CT Exams

A reduction in dose increases the noise in CT scanning, and thus reduces low-contrast detectability (reduces image quality); therefore, in the push to reduce CT dose, one runs the risk of making the scan "non-diagnostic" and of no immediate benefit to the patient, at the cost of a trivial reduction of future cancer risk to the patient. The reason the CT scan was ordered is (ideally) for immediate benefit to the patient, and the risk/benefit ratio very small.

The risk of cancer due to radiation exposure in the diagnostic range is *stochastic* rather than deterministic. If a group of patients are irradiated with a dose in the stochastic range, a small fraction will go on to develop cancer due to chance (i.e., bad luck) while the vast majority of those irradiated will experience no effect at all. In other words, it is the probability that an effect will occur, not the size of the effect, which is proportional to the insult when modeling a stochastic process. However, such complications are not easily detected on an individual basis because most radiation-induced cancers (apart from leukemia) lie latent for at least two decades; and when they do manifest, they are indistinguishable from all other cancers and cannot be reliably attributed to their cause.

The computation of risk vs. dose is now (and probably for evermore) based on the Linear-No-Threshold Theory (LNT), and it is the implications of linearity which are more easily overlooked. Since the slope of the LNT curve is constant, a given dose increment produces the same incremental increase in risk of cancer (Durand 2011; Durand et al. 2012). This means that the first CT scan is just as "dangerous" in terms of absolute cancer risk as the tenth (assuming the same body part is scanned at a similar technique). There is *no buildup of sensitivity* with increasing dose from repeated CT scans. *If there were, the response would not be linear* and *all our current LNT-based risk estimates would then be inapplicable.*

There are those who would argue otherwise, using complex radiobiological arguments; but who also vigorously defend LNT. You can't have it both ways, folks.

10.1.2.1 The Gambler's Fallacy

If one has flipped a coin 20 times and it has come up "heads" every time, then would you bet on "tails" for the next toss? The odds are exactly the same; 1/2 on the twenty-first toss. The cognitive bias leading many to bet on tails, known as "the gambler's fallacy," is also in play in the rush to record cumulative dose in CT. Risks from repeated CT scans is not at all analogous to chopping down a tree, where each axe blow weakens the tree until it finally topples. It is *linear and stochastic* according to the gurus of Radiation Protection.

Cumulative dose estimates are not relevant to rational pre-scan risk versus benefit analysis. If we concede that the relationship between dose and cancer risk is both linear and stochastic, then performing a CT scan is akin to a game of chance. According to the ICRP model (ICRP 2007), the hypothetical risk of a fatal cancer resulting from a typical abdominal scan (8 mSv effective dose) is approximately 0.04% and the implied odds (1:2500) are similar to those of drawing the ace of spades *twice in a row* from a (reshuffled) 52-card deck (1:2703). Likewise, the odds of generating the seed of a fatal cancer are the same for each CT scan, whether it is the first or the tenth scan.

The risk of dying from cancer in one's lifetime from all causes is about 20% (Howlader et al. 2011) while the 8 mSv abdominal scan in this scenario increases this risk by only 0.04% to 20.04%. Even for patients having several previous scans, the risk is still on the order of 100 times less than their natural cancer risk, whereas the scan itself may be life-saving.

10.1.2.2 Death by Coefficient

While the risk coefficient is small (about 5×10^{-5} per mSv), if multiplied by enough people, one can do some serious "killing" with it; as some authors have done "for effect" (to interest the news media). This is usually stated as "this many people will die as a result of CT scans" – neglecting to mention that these are *hypothetical deaths*.

10.2 BOOK SUMMARY

Rigorous phantom dose equations have been derived which also illustrate the significant limitations and common misconceptions concerning the CTDI-paradigm. For example, it does not apply to *shift-variant* scan techniques in which scan parameters are varied during the scan, such as automatic tube current modulation (TCM) in which there is a variation of $mA(z)$. Likewise, it does not apply to stationary table techniques such as perfusion studies using multiple rotations at a fixed z-location or to wide cone beam techniques in which the desired anatomy can be imaged in a single axial rotation without table motion. The equations derived herein (Chapters 7, 8, and 9) for these *shift-variant* and stationary table techniques are rigorous and indicate (in and of themselves) these limitations. Analytic equations are also derived (Chapter 6), based on a scatter kernel of Monte Carlo parentage, which strip away the integral facade of the CTDI-paradigm and provide the reader a better physical understanding of CT dosimetry. That notwithstanding, the CT scanner reports a value of $CTDI_{vol}$ for these techniques based on ad hoc assumptions and "patches" (IEC 2016) in an attempt to extend the life of $CTDI_{vol}$ (and likewise maintain the 100 mm pencil chamber acquisition of its basis, $CTDI_{100}$). The reader is guided herein, as to how to detect (and correct or ignore) the resulting anomalous values of $CTDI_{vol}$ and SSDE. Suggestions are also advanced (or reiterated) for new paradigms of *measurement and phantom dose calculation* based on sound physical principles in lieu of the present ad hoc methodology used for calculation of the scanner-reported $CTDI_{vol}$ and the truncated pencil-chamber measurement methodology. Advanced methods of patient dose calculation such as organ dose calculation often use $CTDI_{vol}$ as a basis as well as the curious parameter $CTDI_{vol}(z)$ which incorrectly assumes that the local dose at z is proportional to the local tube current $mA(z)$; thereby ignoring the basic physics of CT dosimetry, namely that the dose at a point is heavily scatter-dependent (the scatter-to-primary ratio on the central axis of the body phantom is S/P = 13), such that the dose at a given point z depends on $mA(z')$ over the entire scan length. This is self-evident since a considerable dose is deposited by scatter beyond the scan length where the tube current $mA = 0$.

REFERENCES

AAPM Report No. 204, Size-specific dose estimates (SSDE) in pediatric and adult body CT examinations, American Association of Physicists in Medicine, College Park, MD, (2011).

AAPM Report No. 220, Use of water equivalent diameter for calculating patient size and size specific dose estimates (SSDE) in CT, American Association of Physicists in Medicine, College Park, MD, (2014).

Durand D.J., A rational approach to the clinical use of cumulative effective dose estimates. *AJR Am J Roentgenol* 197(1), 160–162, (2011).

Durand D.J., Dixon R.L., and Morin R.L., Utilization strategies for cumulative dose estimates: A review and rational assessment. *J Am Coll Radiol* 9, 480–485, (2012).

Howlader N., Noone A., Krapcho M., Neyman N., Aminou R., Waldron W., Altekruse S.F., Kosary C.L., Ruhl J., Tatalovich Z., Cho H., Mariotto A., Eisner M.P., Lewis D.R., Chen H.S., Feuer E.J., Cronin K.A., and Edwards B.K. (eds)., SEER Cancer Statistics Review, 1975–2008, National Cancer Institute, Bethesda, MD, (2011), http://seer.cancer.gov/csr/1975_2008/, based on November 2010 SEER data submission, posted to the SEER website.

ICRP, Recommendations of the International Commission on Radiological Protection (Users Edition), ICRP Publication 103 (Users Edition), Ann. ICRP 37 (2-4), (2007).

IEC 60601-2-44, *Medical Electrical Equipment — Part 2-44: Particular Requirements for the Basic Safety and Essential Performance of X-ray Equipment for Computed Tomography*, 3rd ed., International Electrotechnical Commission, Geneva, Switzerland, (2016).

Tian X., Li X., Segars W., Dixon R.L., and Samei E., Convolution-based estimation of organ dose in tube current modulated CT. *Phys Med Biol* 6(10) (2016).

Index

Milton Keynes UK
Ingram Content Group UK Ltd.
UKHW052017071024
449327UK00027B/2316